T Level Engineering

T0260422

T Level Engineering is the new technical qualification standing alongside the Academic A Levels, for 16+ students looking to go into engineering. *T Level Engineering* covers the core elements for all the pathways of this qualification.

Whether your sights are set on an engineering university degree, or an advanced apprenticeship, this book covers the essentials needed to get through the 2-year T Level Engineering program. Teachers and work placement managers will like it too as all the sections are broken down into bite-sized pieces – enough for a lesson or two.

You should find *T Level Engineering* easy to understand and readily accessible, even if you have no previous engineering knowledge. The technical terms are explained as they are introduced, and a detailed glossary allows you to check out any specific terms, which is also very useful when writing assignments. You will keep this book handy even after your course has finished and it will provide a reference for a life time.

Andrew Livesey is an experienced Chartered Engineer. He teaches engineering at Ashford College in Kent when not restoring classic MGs, or making electronic gadgets in his workshop. He was a member of the DfE committee responsible for developing the T Levels and is a T Level Ambassador. His Routledge publications include: *Basic Motorsport Engineering* (2011), *Advanced Motorsport Engineering* (2012), *The Repair of Vehicle Bodies* (2018), *Practical Motorsport Engineering* (2018), *Bicycle Engineering and Technology* (2021) and *Motorcycle Engineering* (2021).

T Level Engineering

Technology, Manufacture and Maintenance

Andrew Livesey

Routledge
Taylor & Francis Group

LONDON AND NEW YORK

Cover image: Andrew Livesey

First published 2023
by Routledge
4 Park Square, Milton Park, Abingdon, Oxon OX14 4RN

and by Routledge
605 Third Avenue, New York, NY 10158

Routledge is an imprint of the Taylor & Francis Group, an informa business

© 2023 Andrew Livesey

British Library Cataloguing-in-Publication Data
A catalogue record for this book is available from the British Library

Library of Congress Cataloging-in-Publication Data
Names: Livesey, Andrew, author.
Title: T level engineering : technology, manufacture and maintenance / Andrew Livesey.
Description: Abingdon, Oxon ; New York, NY : Routledge, [2023] | Includes index.
Identifiers: LCCN 2022053416 | ISBN 9781032257501 (pbk) |
ISBN 9781032257518 (hbk) | ISBN 9781003284833 (ebk)
Subjects: LCSH: Engineering–Outlines, syllabi, etc. |
Engineering–Examinations–Study guides. |
Engineers–Great Britain–Certification.
Classification: LCC TA147 .L58 2023 |
DDC 607.60941–dc23/eng/20230208
LC record available at https://lccn.loc.gov/2022053416

ISBN: 978-1-032-25751-8 (hbk)
ISBN: 978-1-032-25750-1 (pbk)
ISBN: 978-1-003-28483-3 (ebk)

DOI: 10.1201/9781003284833

Typeset in Sabon
by Newgen Publishing UK

de Livesey

This book is dedicated to the memory of Sir Michael Livesey 1614 to 1665. Lord Lieutenant of Kent and Member of Parliament. His actions helped to found the modern parliamentary system, a system copied by many other countries across the world.

Contents

    Procedures   195

13 Health and Safety – Standards, Acts, Legislation and
    Risk Analysis   199

14 Business, Commercial and Financial Awareness   209

15 Professional Responsibilities, Attitudes and Behaviours   219

16 Stock and Asset Management   223

17 Quality Assurance, Control and Improvement   226

18 Principles and Practices of Continuous Improvement   229

19 Project Management Principles, Techniques and
    Practices   233

20 My Project   237

    Glossary   240
    Index   242

Preface

The introduction of the T Level Qualifications is the key to changing the face of education, educating the new generation of technical people, the people the world needs to generate a bright and prosperous future.

T for Technical, this is very important, the world needs people who can actually do something technical and practical. T Levels lead into apprenticeships, including degree apprenticeships, and higher education programmes.

I'm proud to have been a member of one of the T Level committees which wrote the content for the Engineering T Levels. This book sets out to cover the key topics for all of the Engineering T Levels, it is designed to fit in the college bag of every learner; then to be kept on the shelf for reference. Please write in it, add notes, do calculations and make sketches in it, as well as reading it.

Enjoy your learning and consequent career, I'm enjoying mine.

Special thank you to colleagues at Ferrari Europe and Maserati GB.

Eur Ing Andrew Livesey MA CEng MIMechE BEd (Hons) AAE FIMI DipILM
Andrew@Livesey.US
Herne Bay
Kent

Abbreviations and Symbols

The abbreviations are generally defined by being written in full when the relevant technical term is first used in the book. In a very small number of cases, an abbreviation may be used for two separate purposes, usually because the general concept is the same; but the use of a superscript or subscript would be unnecessarily cumbersome; in these cases, the definition should be clear from the context of the abbreviation. The units used are those of the internationally accepted *System International* (SI). However, because of the large American input, and the desire to retain the well-known Imperial system of units by many engineers, where appropriate Imperial equivalents of SI units are given. Therefore, the following is intended to be useful for reference only and is neither exhaustive nor definitive.

Greek Alphabet Symbols

A	(alpha) angle
Λ	(lambda) angle of inclination
M	(mu) co-efficient of friction
Ω	(omega) rotational velocity
P	(rho) air density
H	(eta) efficiency
Θ	(theta) angle

General Abbreviations

a	acceleration
A	Ampere
ABS	acrylonitrile butadiene styrene (a plastic)
AC	alternating current
AF	across flats – bolt head size

AFFF	aqueous film-forming foam (fire-fighting)
bar	atmospheric pressure – 101.3 kPa or 14.7 psi as standard or normal
BATNEEC	best available technique not enabling excessive cost
BS	British Standard
BSI	British Standards Institute
C	Celsius; or Centigrade
CAD	computer-aided design
CAE	computer-aided engineering
CAM	computer-aided manufacturing
C_D	aerodynamic co-efficient of drag
CG	centre of gravity, also CoG
CIM	computer-integrated manufacturing
C_L	aerodynamic coefficient of lift
cm	centimetre
cm^3	cubic centimetres – capacity; also called cc. 1000 cc is 1 litre
CO	carbon monoxide
CO_2	carbon dioxide
COSHH	Control of Substances Hazardous to Health (Regulations)
CP	centre of pressure
CR	compression ratio
D	diameter
d	distance
dB	decibel (noise measurement)
DC	direct current
deg	degree (angle or temperature), also °
dia.	diameter
DTI	dial test indicator
EC	European Community
ECU	electronic control unit
EFI	electronic fuel injection
EN	European Norm – European Standard
EPA	Environmental Protection Act; or Environmental Protection Agency
EU	European Union
f	frequency
F	Fahrenheit, force
ft	foot
ft/min	feet per minute
g	gravity; or gram
gal	gallon (US gallon is 0.8 of UK gallon)
GRP	glass-reinforced plastic
HASAWA	Health and Safety at Work Act

HGV	heavy goods vehicle (used also to mean LGV – large goods vehicle)
hp	horse power (CV in French, PS in German)
HSE	Health and Safety Executive; also, health, safety and environment
HT	high tension
I	inertia
ID	internal diameter
IMechE	Institution of Mechanical Engineers
IMI	Institute of the Motor Industry
in^3	cubic inches – measure of capacity
IR	infra-red
ISO	International Standards Organization
k	radius of gyration
kph	kilometres per hour
l	length
L	wheelbase
LH	left hand
LHD	left-hand drive
LHThd	left-hand thread
LPG	liquid petroleum gas
lumen	light energy radiated per second per unit solid angle by a uniform point source of 1 candela intensity
lux	unit of illumination equal to 1 lumen/m^2
M	mass
MAX	maximum
MIG	metal inert gas (welding)
MIN	minimum
N	Newton; or normal force
Nm	Newton metre (torque)
No	number
OD	outside diameter
OL	overall length
OW	overall width
P	power, pressure or effort
Part no	part number
PPE	personal protective equipment
pt	pint (UK 20 fluid ounces, USA 16 fluid ounces)
PVA	polyvinyl acetate
PVC	polyvinyl chloride
Q	heat energy
r	radius
R	reaction
Ref	reference

RH	right hand
rpm	revolutions per minute; also RPM and rev/min
RTA	Road Traffic Act
RWD	rear-wheel drive
std	standard
STP	standard temperature and pressure
TE	tractive effort
TIG	tungsten inert gas (welding)
V	velocity; or volt
VOC	volatile organic compounds
W	weight
w	width
WB	wheel base
x	longitudinal axis of vehicle or forward direction
y	lateral direction (out of right side of vehicle)
z	vertical direction relative to vehicle

Superscripts and subscripts are used to differentiate specific concepts.

SI Units

cm	centimetre
K	Kelvin (absolute temperature)
kg	kilogram (approx. 2.25 lb)
km	kilometre (approx. 0.625 mile or 1 mile is approx. 1.6 km)
kPa	kilopascal (100 kPa is approx. 15 psi, that is atmospheric pressure of 1 bar)
kV	kilovolt
kW	kilowatt
l	litre (approx. 1.7 pint)
l/100 km	litres per 100 kilometres (fuel consumption)
m	metre (approx. 39 inches)
mg	milligram
ml	millilitre
mm	millimetre (1 inch is approx. 25 mm)
N	Newton (unit of force)
Pa	Pascal
ug	microgram

Imperial Units

ft	foot (= 12 inches)
hp	horse power (33,000 ftlb/minute; approx. 746 Watt)
in	inch (approx. 25 mm)
lb/in^2	pressure, sometimes written psi
lbft	torque (10 lbft is approx. 13.5 Nm)

Chapter 1

Engineering Development and Innovation

New developments in engineering happen daily; new innovations are developed on the workshop floor as well as in the engineering design and drawing offices. This may be something as simple as moving a fixing a millimetre, completely redesigning a component or inventing something completely new.

You might develop a list of new products and innovations which have come into common use in the past few years. Many of these relate to microelectronics and mobile communications, these items need a high level of understanding of materials and mathematics. However, there many much simpler-to-grasp concepts which are of great value to the people who need them, and still in use many years later.

Breaking Down Large Items to Build Them

Lake Titicaca is 190 km long and 3,812 m high in the Andes mountains of South America. In 1892, they wanted a ship to move goods and people across the lake. Although they had small boats, they wanted a large ship for the growing trade. The area around Dumbarton on the River Clyde in Scotland is well known for building good ships. Now imagine yourself as an engineer in Dumbarton, probably not even knowing where Lake Titicaca is, being asked to design a ship for use on this lake. The added requirement is that it has to be transported there from Scotland, and it cannot be sailed there, as the lake is high in the mountains up a narrow track.

They designed the ship SS Coya, weight of 546 tonnes, and 52 metres (170 feet) long in 1892, the engineers broke the design down into parts small enough to be moved by mules, and assembled it by the side of the lake. SS Coya is still in use on Lake Titicaca.

This technology of designing items into small parts has led to being able to buy equipment and furniture in what we now call flat-pack. Easy to transport and easy to assemble on site. An engineer/architect called Peter Huf builds flat-pack houses, known as Huf Haus, which are transported to

DOI: 10.1201/9781003284833-1

Figure 1.1 Latest Maserati range.

the building site by lorry ready for assembly – it takes about two weeks to assemble one.

Standardisation

Up to the middle of the 19th century all engineered items were individually made. For example, each link of a chain was individually forged, each thread of a bolt and nut was cut by hand. So, although they were similar in size, they were not identical. The American Civil War of 1861–1865 led to a great demand for guns and rifles. At this time they were made individually, by hand, each having a slightly different diameter barrel dependent on the individual maker, and supplied with a mould for making the shot to suit. Of course, this meant that they could not be readily interchanged between soldiers on the battlefield. A production method was developed so that all the rifles had the same diameter barrel, thereby they were fully interchangeable. This was done using a production line system, all the barrels were bored on the same machine. About two million rifles were made during this period.

Subsequently, Henry Ford developed the production line method for making cars with interchangeable parts and speeding up car production. It is still in use in most manufacturing processes.

Time line – the dates given are those recognised by most sources to give the reader an idea of the activities.

Date	Engineering Achievement	World Events
1436	Printing press invented by Johan Gutenberg	
1698	Steam engine – used to pump water from mines by Thomas Savery	1760–1830 Industrial Revolution mass manufacturing changes based on use of steam power
1752	Electricity – discovered by Benjamin Franklin	
1779	Iron Bridge – first all-metal bridge built in Shropshire	
1804	First steam railway – Richard Trevithick builds first steam railway	
1818	Draisienne bicycle – built in Germany	
1822	Charles Babbage Computer	
1825	Stockton and Darlington Railway opened	
1825	Aluminium discovered	
1832	First electric car built	
1843	SS Great Britain built by Brunel	
1856	Aluminium first used	
1861	Nicolaus Otto builds atmospheric engine	1861–1865 American Civil War, guns produced in millions
1865	Mass production of bicycles starts	
1876	Four-stroke petrol engine built by Otto	
1876	Telephone invented by Alexander Graham Bell	
1877	Ball and roller bearings	
1878	Thomas Edison invents the light bulb, cinematography and the gramophone	
1880	Renold roller chain, started use of chains on bicycles	
1885	First motor car and motorcycle invented	
1895	X ray	
1897	Seamless butted tubing – by Reynolds	
1903	Flight – airplane	1899–1902 Boer War in South Africa
1913	Stainless steel discovered by Harry Brearley	
1913	Traffic lights first used in Paris	
1926	Television – invented by John Logie Baird	
1927	Refrigerator – invented by Fred Wolf	1914–1918 World War 1, engineering production turned to military vehicles and weapons
1931	Jet engine – invented by Sir Frank Whittle	
1938	Glass fibre – thin glass tubes set in resin	
1941	Robotics first used	
1960	Fibre optic – transmits both light and conducts electricity	
1963	Carbon fibre – both ultra-light and ultra-strong	
1965	Post Office Tower – building with top part that turns around	1939–1944 World War 2, worldwide production of aeroplanes was about one million
1973	First PC computer	
1980	3D printing first used	
1983	Internet invented	
1984	Motorola make first mobile phone	
2013	Shard built – a skyscraper built largely of glass	
2021	Autonomous cars allowed on roads	

Figure 1.2 Vintage MG.

If you study the history of engineering you will find that several people invented the same thing at about the same time. The changes which are taking place every day go largely unnoticed, they are often unseen; but they may have a very big effect on how we do things.

AI – artificial intelligence – uses a stimulus–response logic system to carry out tasks or answer questions. The actual task performed will depend on the questions and answers. For example, if you telephone your bank or credit card company:

You	AI telephone system
Dial the bank	Are you an existing customer?
Yes	Enter your account number?
No	Do you want to open an account?
Enter account number	Which service do you require – followed by list of options?

AI can perform complex tasks using multiple inputs, based on their functionality, the three current main types are:

> **Reactive machine** – This AI has no memory power and does not have the ability to learn from past actions.

Limited theory machine – With the addition of memory, this AI uses past information to make better decisions.

Artificial narrow intelligence – ANI – A system that performs narrowly defined programmed tasks. This AI has a combination of reactive and limited memory. Most of today's AI applications are in this category.

Speed of processing data – One of the important advantages of using AI is the speed at which data can be processed. In other words, how quickly calculations can be made, or questions processed. Currently the fastest speed is 10 trillion operations per second – TeraOPs/s.

Tech note

Take a trillion as a thousand million. So high-speed data processing is written out in numbers as 10,000,000,000 calculations per second. The human brain speed is about 60 bits per second – a bit being taken as a single calculation.

An everyday example of AI is the satellite navigation systems used in cars and on motorcycles and cycles.

It takes inputs from: your current position, where you want to go to, your preferred choice of roads, how quickly you want to get there. It compares this to which roads are available, weather conditions, road works, accidents or hold-ups ahead, traffic conditions. It then suggests a route, and gives an estimated time for the journey. Should any of the data change as your journey progresses, it will offer you an alternative route.

Robotics – Robots are widely used in manufacturing and similar controlled environments. The robot is usually pre-programmed to carry out a set routine. That is a routine of movements to complete a task. Lasers are often used for the robot to detect the work piece.

Smart materials – These have properties that react to changes in their environment. This means that one of their properties can be changed by an external condition, such as temperature, light, pressure, electricity, voltage, pH or chemical compounds. This change is reversible and can be repeated many times. There is a wide range of different smart materials. Each offers different properties that can be changed. Some materials are very good and cover a huge range of the scales.

Some examples are:

- **Piezoelectric materials** are materials that produce a voltage when stress is applied. Since this effect also applies in a reverse manner, a voltage across the sample will produce stress within the sample. Suitably designed structures made from these materials can, therefore, be made that bend, expand or contract when a voltage is applied.

- **Shape-memory alloys and shape-memory polymers** are materials in which large deformation can be induced and recovered through temperature changes or stress changes.
- **Photovoltaic materials** convert light to electrical current.
- **Electroactive polymers** change their volume by voltage or electric fields.
- **Magnetostrictive** materials change in shape under the influence of a magnetic field and also exhibit a change in their magnetization under the influence of mechanical stress.
- **Temperature-responsive** polymers are materials which undergo changes with variations in temperature.
- **Photochromic materials** change colour in response to light, for example, light-sensitive sunglasses that darken when exposed to bright sunlight.
- **Ferro fluids** are magnetic fluids which are affected by magnets and magnetic fields.
- **Self-healing** materials have the intrinsic ability to repair damage due to normal usage, thus expanding the material's lifetime. Vehicle paint which is both self-healing and polychromatic is sometimes used on custom vehicles.
- **Thermoelectric** materials are used to build devices that convert temperature differences into electricity and vice versa.

  Hybrid technology – Hybrid technology is increasingly used for motor vehicles to make them capable of high-mileage journeys and low emissions too. That is, they have both an internal combustion engine and an electric motor. The use of two separate power plants gives outstanding acceleration also.

  Autonomous systems – Many factories and warehouses use autonomous vehicles – ones without drivers. Autonomous, or self-driving cars are now allowed on UK roads.

  Sustainability – This is a societal goal that broadly aims for humans to safely co-exist on planet Earth over a long time. Specific definitions of sustainability are difficult to agree on and therefore vary in the literature and over time. Sustainability is commonly described along the lines of three dimensions: the environmental, economic and social dimensions. This concept can be used to guide decisions at the global, national and individual levels. In engineering one of the biggest problems is wastage of raw materials. Therefore design, manufacture and repair activities are usually carried out using the minimal amount of materials needed. This, of course, also reduces costs.

  Product life cycle – Generally, most engineered products do not last a long time. Therefore manufacturers must state how these can

Figure 1.3 Apartments in Florida.

be dealt with at the end of their useful life. Vehicle breakers must follow an approved procedure including disposal of engine oil and brake fluid.

Changing technology – Technology, how we do things, is continually changing. We need to do this with reference to sustainability and overall safety.

Alternatives – In engineering there are generally several ways to carry out any job in manufacturing, installation, repair and maintenance. It is important to keep up to date with alternatives. Of course, this includes sustainability and changing technology.

Figure 1.4 Chilli Folding Electric Bicycle.

Waste and disposal – This is now largely controlled by legislation. General waste and recyclable materials are generally handled by the local authority, or an approved company. In engineering there is usually both scrap metal and contaminated liquids; if there is sufficient quantity these can usually be sold for recycling.

Skills and Questions

Engineering development and innovation comes with a mixture of experience and sudden flashes of insight. Whether you are designing something, repairing something or installing something, best practice is to see what someone did before, look at any problems or faults, then work out how to do it better.

Keeping up to date is part of the requirements for a professional engineer. A good way to keep up to date is to join a professional institutions. Student membership is usually free of charge. The suggestion is that you choose and join one of the professional institutions, and then you will usually get a magazine, updating emails and invitations to meetings.

Chapter 2

Introduction to Workplace Health and Safety Procedures and Practices

Health and Safety, and the environment, are controlled by a number of Acts of Parliament and subsequent regulations and statutory instruments. These may have regional variations, or specifics relating to the various parts of the engineering industry. These topics are also impacted by other laws, for instance those relating to the environment and the countryside. Perhaps the most important aspect of the Health and Safety laws is that they are **vicarious**. That is, if you are a manager, or supervisor, of an engineering company, you may be prosecuted for any injury or death caused by one of the technicians or other staff actions, as well as the staff involved being prosecuted. Breaking Health and Safety, or environmental laws may result in custodial sentences as well as fines and damages to the injured parties. Meeting the requirements of the Health and Safety, and environmental laws and regulations is the responsibility of everybody – ignorance of the law is not an excuse, so you need to take care.

Personal Health and Safety Procedures

Skin care (personal hygiene) systems – employees should be aware of the importance of personal hygiene and should follow correct procedures to clean and protect their skin in order to avoid irritants causing skin infections and dermatitis. All personnel should use a suitable barrier cream before starting work and again when recommencing work after a break. There are waterless hand cleaners available which will remove heavy dirt on skin prior to thorough washing. When the skin has been washed, after-work creams will help to restore its natural moisture.

Hand protection – engineers and technicians are constantly handling substances which are harmful to health. The harmful effect of liquids, chemicals and materials on the hands can be prevented, in many cases, by wearing the correct type of gloves.

Protective clothing is worn to protect the worker and their clothes from coming into contact with dirt, extremes of temperature, falling objects and

DOI: 10.1201/9781003284833-2

chemical substances. The most common form of protective clothing for the engineer is a coat made from good-quality cotton, preferably flame-proof. Worn and torn materials should be avoided as they can catch in moving machinery. Where it is necessary to protect the skin, closely fitted sleeves should be worn down to the wrist with the cuffs fastened. All overall buttons must be kept fastened, and any loose items such as ties and scarves should not be worn.

Head protection is very important to the engineering worker when working underneath an object. A lightweight safety helmet, normally made from aluminium, fibreglass or plastic, should be worn if there is any danger from falling objects, and will protect the head from damage when working in a confined space. Hats and other forms of fabric headwear keep out dust, dirt and overspray and also prevent long hair (tied back) becoming entangled in moving equipment.

Eye protection is required when there is a possibility of eye injury from flying particles when using a grinder disc sander, power drill or pneumatic chisel, or other machinery. Many employers now require all employees to wear some form of eye protection when they are in the workshop.

- **Lightweight safety spectacles** with adjustable arms and with side shields for extra protection. There is a choice of impact grades for the lenses.
- **General-purpose safety goggles** with a moulded PVC frame which is resistant to oils, chemicals and water. These have either a clear acetate or a polycarbonate lens with BS impact grades 1 and 2.
- **Face shields** with an adjustable head harness and deep polycarbonate brow guard with replaceable swivel-up clear or anti-glare polycarbonate visor BS grade 1, which gives protection against sparks, molten metal and chemicals.
- **Welding helmet or welding goggles** with appropriate shaded lens to BS regulations. These must be worn at all times when welding. They will protect the eyes and face from flying molten particles of steel when gas welding and brazing, and from the harmful light rays generated by the arc when MIG/MAG, TIG or MMA welding

Foot protection – Safety footwear is essential in the workshop environment. Boots or shoes with steel toecaps will protect the toes from falling objects, the high sides of boots protect the ankles. Rubber boots will give protection from acids or in wet conditions. Never wear defective footwear as this becomes a hazard in any workshop environment.

Ear protection – the **Noise at Work Regulations** define three action levels for exposure to noise at work:

- A daily personal exposure of up to 80 dB. Where exposure exceeds this level, suitable hearing protection must be provided on request.

Figure 2.1 Lightweight safety glasses.

Figure 2.2 Face protection mask.

- A daily personal exposure of up to 85 dB. Above this second level of provision, hearing protection is mandatory.
- A peak sound pressure of 87 dB.

Tech note

Noise is measured in decibels (dB) – this is measured on a scale based on logarithms. That is to say that increases do not follow the normal arithmetic scale in terms of increase in noise. An increase of 3 dB, from say 84 dB to 87 dB will give a doubling of the noise heard. So, to cut the reading by 3 dB will reduce the noise heard by half.

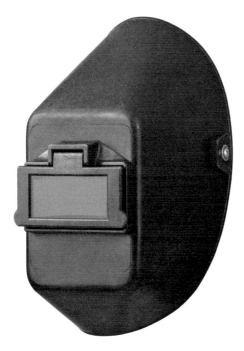

Figure 2.3 Automatic welding mask.

Where the second or third levels are reached, employers must designate ear protection zones and require all who enter these zones to wear ear protection. Where the third level is exceeded, steps must be taken to reduce noise levels as far as is reasonably practicable. In every case where there is a risk of significant exposure to noise, assessment must be carried out and action taken to minimise hearing damage.

The first two noise action levels relate to exposure over a period (one day) and are intended to cater for the risks of prolonged work in noisy surroundings. The third level is related to sudden impact noises like those occurring in metal-working procedures.

Fire Precautions

The Fire Precautions (workplace) Regulations 1992/1993/1997 replaced and extended the old Fire Precautions Act 1971. These Regulations are aligned with standard practice in EC Directives in placing the responsibility for compliance on employers. They require employers not only to assess risks from fire, but to include the preparation of an evacuation plan, to train staff in fire precautions, and to keep records. Workplaces with fewer than 20 employees may require emergency lighting points and fire warning systems. The self-employed who do not employ anyone but whose premises are regularly open to the public may only require fire extinguishers and warning signs; they will, however, need to be able to demonstrate that there is a means of escape in case of fire. Where five or more persons work on the premises as employees, all assessments need to be recorded in writing.

Most of these requirements were already covered by existing legislation. There is a detailed requirement for the recording of assessments, the provision of training, and the requirement that means of fighting fire, detecting fire and giving warning in case of fire, be maintained in good working order.

What is fire? Fire is a chemical reaction called combustion (usually oxidation resulting in the release of heat and light). To initiate and maintain this chemical reaction, or in other words for an outbreak of fire to occur and continue, the following elements are essential

FUEL: A combination substance, solid, liquid or gas.
OXYGEN: Usually air, which contains 21 per cent oxygen.
HEAT: The attainment of a certain temperature – once a fire has started it normally maintains its own heat supply.

Methods of extinguishing of fire, because three ingredients are necessary for fire to occur, it follows logically that if one or more of these ingredients are removed, fire will be extinguished. Basically three methods are employed to extinguish a fire: removal of heat (cooling); removal of fuel (starving); and removal or limitation of oxygen (blanketing or smothering).

Removal of heat – if the rate of heat generation is less than the rate of dissipation, combustion cannot continue. For example, if cooling water can

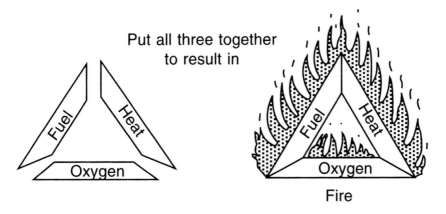

Put all three together
to result in

Fuel   Heat
Oxygen

Fuel   Heat
Oxygen

Fire

Figure 2.4 The fire triangle.

absorb heat to a point where more heat is being absorbed than generated, the fire will go out.

Removal of fuel – this is not a method that can be applied to fire extinguishers. The subdividing of risks can starve a fire, prevent large losses and enable portable extinguishers to retain control; for example, part of a building may be demolished to provide a fire stop. The following advice can contribute to a company's fire protection programme:

* What can cause fire in this location, and how can it be prevented?
* If fire starts, regardless of cause, can it spread?
* If so, where to?
* Can anything be divided or moved to prevent such spread?

Removal or limitation of oxygen – it is not necessary to prevent the contact of oxygen with the heated fuel to achieve extinguishment. It will be found that where most flammable liquids are concerned, reducing the oxygen in the air from 21 to 15 per cent or less will extinguish the fire. Combustion becomes impossible even though a considerable proportion of oxygen remains in the atmosphere. This rule applies to most solid fuels although the degree to which oxygen content must be reduced may vary. Where solid materials are involved they may continue to burn or smoulder until the oxygen in the air is reduced to 6 per cent. There are also substances which carry within their own structures sufficient oxygen to sustain combustion.

Fire Risks in the Workshop

Fire risks in the workshop cover all classes of fire: class A, is paper, wood and cloth; class B, is flammable liquids such as oils, spirits, alcohols, solvents and grease; class C, is flammable gases such as acetylene, propane, butane; and also electrical risks. It is essential that fire is detected and extinguished in the early stages. Workshop staff must know the risks involved and should be aware of the procedures necessary to combat fire. Engineering personnel should be aware of the various classes of fire and how they relate to common workshop practice.

Class A Fires: Wood, Paper and Cloth

Today wood is not generally used in cars, although there are exceptions. Cloth materials are used for some main trim items and are therefore a potential fire hazard. The paper used for masking purposes is a prime area of concern. Once it has done its job and is covered in overspray it is important that it is correctly disposed of, ideally in a metal container with a lid, and not scrunched up and thrown on the floor to form the potential start of a deep-seated fire.

Class B Fires: Flammable Liquids

Flammable liquids are the stock materials used in engineering refinishing processes: gun cleaner to clear finish coats, cellulose to the more modern finishes, can all burn and produce acrid smoke.

Class C Fires: Gases

An increasing percentage of cars run on liquid gas (LPG/LNG), also welding gases or propane space heaters not only burn but can be the source of ignition for class A or B fires.

Electrical Hazards

Electricity is not of itself a class of fire. It is, however, a potential source of ignition for all of the fire classes mentioned above. The Electricity at Work Regulations cover the care of cables, plugs and wiring. In addition, in the workshop the use of welding and cutting equipment produces sparks which can, in the absence of good housekeeping, start a big fire. Training in how to use firefighting equipment can stop a fire in its early stages. Another hazard is the electrical energy present in all car batteries. A short-circuit across the terminals of a battery can produce sufficient energy to form a weld and in turn heating, a prime source of ignition. When tackling a car fire a firefighter will always try to disconnect the battery, as otherwise any attempt to extinguish a fire can result in the re-ignition of flammable vapours.

General Precautions to Reduce Fire Risk

- Good housekeeping means putting rubbish away rather than letting it accumulate.
- Read the manufacturer's material safety data sheets so that the dangers of flammable liquids are known.
- Only take from the stores sufficient flammable material for the job in hand.
- Materials left over from a specific job should be put back into a labelled container so that not only you but anyone (and this may be a firefighter) can tell what the potential risk may be.
- Take care when welding that sparks or burning paint coating does not cause a problem, especially when working in a confined area.

The keys to fire safety are:

- Take care.
- Think.

- Train staff in the correct procedures before things go wrong.
- Ensure that these procedures are written down, understood and followed by all personnel within the workshop.

Tech note

Any carbon-based material will burn in air if at the temperature needed for combustion – be aware of this with dust in the workshop.

Types of Portable Fire Extinguishers

Water is the most widely used extinguisher agent. With portable extinguishers, a limited quantity of water can be expelled under pressure and its direction controlled by a nozzle.

There are basically two types of water extinguishers. The gas (CO_2) cartridge-operated extinguisher, when pierced by a plunger, pressurises the body of the extinguisher, thus expelling the water and producing a powerful jet capable of rapidly extinguishing class A fires. In stored pressure extinguishers, the main body is constantly under pressure from dry air or nitrogen, and the extinguisher is operated by opening the squeeze grip discharge valve. These extinguishers are available with 6-litre or 9-litre capacity bodies and thus provide alternatives of weight and accessibility.

Foam is an agent most suitable for dealing with flammable liquid fires. Foam is produced when a solution of foam liquid and water is expelled under pressure through a foam-making branch pipe at which point air is entrained, converting the solution into foam. Foam extinguishers can be pressurised either by a CO_2 gas cartridge or by stored pressure. The standard capacities are 6 and 9 litres.

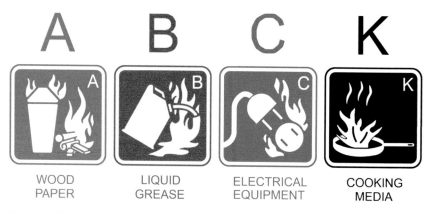

Figure 2.5 Types of fires – A, B, C and K each require a different type of fire extinguisher.

Types of fire extinguisher

TYPE	Class A	Class B	Class C	Class D	Electrical	Class F	Comments
	Combustible materials (e.g. paper & wood)	Flammable liquids (e.g. paint & petrol)	Flammable gases (e.g. butane & methane)	Flammable metals (e.g. lithium & potassium)	Elecrical equipment (e.g. computers & generators)	Deep fat fryers (e.g. chip pans)	
WATER	O	X	X	X	X	X	Do not use on liquid or electric
FOAM	O	O	X	X	X	X	Not suited to domestic use
DRY POWDER	O	O	O	O	O	X	Can be used safely up to 1000 volts
CO₂	X	O	X	X	O	X	Safe on both high and low voltage
WET CHEMICAL	O	X	X	X	X	O	Use on extremely high temperatures

Figure 2.6 Uses of different types of fire extinguishers.

Spray foam. Unlike conventional foams, aqueous film-forming foam (AFFF) does not require to be fully aspirated in order to extinguish fires. Spray foam extinguishers expel an AFFF solution in an atomised form which is suitable for use on class A and class B fires. AFFF is a fast and powerful means of tackling a fire and seals the surfaces of the material, preventing re-ignition. The capacity can be 6 or 9 litres, and operation can be by CO_2 cartridge or stored pressure.

Carbon Dioxide

Designed specifically to deal with class B, class C and electrical fire risks, these extinguishers deliver a powerful concentration of carbon dioxide gas under great pressure. This not only smothers the fire very rapidly, but is also non-toxic and is harmless to most delicate mechanisms and materials.

Dry Powder

This type of extinguisher is highly effective against flammable gases, open or running fires involving flammable liquids such as oils, spirits, alcohols, solvents and waxes, and electrical risks. The powder is contained in the metal body of the extinguisher from which it is supplied either by a sealed gas cartridge, or by dry air or nitrogen stored under pressure in the body of the extinguisher in contact with the powder.

Dry powder extinguishers are usually made in sizes containing 1–9 kg of either standard powder or (preferably and more generally) all-purpose powder, which is suitable for mixed-risk areas.

Choosing and Siting Portable Extinguishers

Because there is such a variety of fire risks in workshops, it is important to analyse these risks separately and (with the help of experts such as fire officers) to choose the correct firefighting medium to deal with each possible fire situation. It should be noted that portable fire extinguishers are classified as first-aid firefighting and are designed for ease of operation in an emergency. It is important to realise that because they are portable they have only a limited discharge. Therefore their siting, together with an appreciation of their individual characteristics, is fundamental to their success in fighting fire.

Safety Signs in the Workshop

It is a legal requirement that all safety signs used in a workshop comply with BS 5378: Part 1. Each of these signs is a combination of colour and design, within which the symbol is inserted. If additional information is required, supplementary text may be used in conjunction with the relevant symbol, provided that it does not interfere with the symbol. The text can be in an oblong or square box of the same colour as the sign, with the text in the relevant contrasting colour, or the box can be white and the text black. BS 5378 divides signs into four categories, these are:

Prohibition signs have a red circular outline and crossbar running from top left to bottom right on a white background. The symbol displayed on the sign must be black and placed centrally on the background, without obliterating the crossbar. The colour red is associated with 'stop' or 'do not'.

Warning signs have a yellow triangle with a black outline. The symbol or text used on the sign must be black and placed centrally on the background. This combination of black and yellow identifies caution.

Mandatory signs have a blue circular background. The symbol or text used must be white and placed centrally on the background. Mandatory signs indicate that a specific course of action is to be taken.

Class of fire	Water	Foam (AFFF)	CO₂ gas	Powder
A Paper Wood Textile Fabric	✓	✓		✓
B Flammable liquids		✓	✓	✓
C Flammable gases			✓	✓
⚡ Electrical hazards			✓	✓
🚗 Vehicle protection		✓		✓

Figure 2.7 Choice of fire extinguishers.

Key to British and European Standard safety signs

| Prohibition
Don't do | Warning
Risk of danger | Safe condition
The safe way | Mandatory
Must do |

Figure 2.8 Standard safety signs.

Safe condition signs provide information for a particular facility and have a green square or rectangular background to accommodate the symbol or text, which must be in white. The safety colour green indicates 'access' or 'permission'.

Fire safety signs are specified by BS 5499, which gives the characteristics of signs for fire equipment, precautions and means of escape in case of fire. It uses the basic framework concerning safety colours and design adopted by BS 5378.

General Safety Precautions in the Workshop

The Health and Safety Act is designed to ensure that:

- Employers provide a safe working environment with safety equipment and appropriate training.
- Employees work in a safe manner using the equipment provided and follow the guidance and training which is provided.
- Customers and others entering any business premises are safe and protected.

The following section looks at some of the details related to health and safety, in all cases you should ask yourself the following questions:

- Are there any regulations relating to this, what are they, and what do I need to do to follow them?
- What is the risk involved and how can I remove, or reduce, the risk?
- Is any documentation needed?

Remember

It is all about keeping yourself, your colleagues and your customers' safe, as you would want them to keep you safe too.

Particular hazards may be encountered in the workshop, and safety precautions associated with them are as follows:

1 Do wash before eating, drinking or using toilet facilities to avoid transferring the residues of sealers, pigments, solvents, filing of steel, lead and other metals from the hands to the inner parts and other sensitive areas of the body.

2 Do not use kerosene, thinners or solvents to wash the skin. They remove the skin's natural protective oils and can cause dryness and irritation or have serious toxic effects.

3 Do not overuse waterless hand cleaners, soaps or detergents, as they can remove the skin's protective barrier oils.

4 Always use barrier cream to protect the hands, especially against fuels, oils, greases, hydrocarbon solvents and solvent-based sealers.

5 Do follow work practices that minimise the contact of exposed skin and the length of time liquids or substances stay on the skin.

6 Do thoroughly wash contaminants such as used engine oil from the skin as soon as possible with soap and water. A waterless hand cleaner can be used when soap and water are not available. Always apply skin cream after using waterless hand cleaner.

7 Do not put contaminated or oily rags in pockets or tuck them under a belt, as this can cause continuous skin contact.

8 Do not dispose of dangerous fluids by pouring them on the ground, or down drains or sewers.

9 Do not continue to wear overalls which have become badly soiled or which have acid, oil, grease, fuel or toxic solvents spilt over them. The effect of prolonged contact from heavily soiled overalls with the skin can be cumulative and life-threatening. If the soilants are, or become, flammable from the effect of body temperature, a spark from welding or grinding could envelop the wearer in flames with disastrous consequences.

10 Do not clean dusty overalls with an airline: it is more likely to blow the dust into the skin, with possible serious or even fatal results.

11 Do wash contaminated or oily clothing before wearing it again.

12 Do discard contaminated shoes.

13 Wear only shoes which afford adequate protection to the feet from the effect of dropping tools and sharp and/or heavy objects on them, and also from red hot and burning materials. Sharp or hot objects could easily penetrate unsuitable footwear such as canvas plimsolls or trainers. The soles of the shoes should also be maintained in good condition to guard against upward penetration by sharp or hot pieces of metal.

14 Ensure gloves are free from holes and are clean on the inside. Always wear them when handling materials of a hazardous or toxic nature.

15 Keep goggles clean and in good condition. The front of the glasses or eyepieces can become obscured by welding spatter adhering to them. Renew the glass or goggles as necessary. Never use goggles with cracked glasses.

16 Always wear goggles when using a bench grindstone or portable grinders, disc sanders, power saws and chisels.

17 When welding, always wear adequate eye protection for the process being used. MIG/MAG welding is particularly high in ultraviolet radiation which can seriously affect the eyes.

18 Glasses, when worn, should have 'safety' or 'splinter-proof' glass or plastic lenses.

19 Electric shock can result from the use of faulty and poorly maintained electrical equipment or misuse of equipment. All electrical equipment must be frequently checked and maintained in good condition. Flexes, cables and plugs must not be frayed, cracked, cut or damaged in any way. Equipment must be protected by the correctly rated fuse.

20 Use low-voltage equipment wherever possible (110 volts).

21 In case of electric shock:
    (a) Avoid physical contact with the victim.
    (b) Switch off the electricity.
    (c) If this is not possible, drag or push the victim away from the source of the electricity using non-conductive material.
    (d) Commence resuscitation if trained to do so.
    (e) Summon medical assistance as soon as possible.

Electrical Hazards

The Electricity at Work Act 1989 fully covers the responsibilities of both the employee and the employer. As an engineer you are obliged to follow these regulations for the protection of yourself and your colleagues. Some of the important points to be aware of are given below.

Voltages – the normal mains electricity voltage via a three-pin socket outlet is 240 volts; heavy-duty equipment such as workshop hoists uses 415 volts in the form of a three-phase supply. Both 240-volt and 415-volt supplies are likely to kill anybody who touches them. Supplies of 415 volts must be used through a professionally installed system. If 240 volts is used for power tools, then a safety circuit breaker should be used. A safer supply for power tools is 110 volts; this may be wired into the workshop as a separate circuit or provided through a safety transformer. Inspection hand-lamps are safest with a 12-volt supply; but for reduced current flow 50-volt hand-lamp systems are frequently used.

Checklist

Before using electrical equipment the engineer is advised to check the following:

1 Cable condition – check for fraying, cuts or bare wires.

2 Fuse rating – the fuse rating should be correct for the purpose as recommended by the equipment manufacturer.

3   Earth connection – all power tools must have sound earth connections.
4   Plugs and sockets – do not overload plugs and sockets; ensure that only one plug is used in one socket.
5   Water – do not use any electrical equipment in any wet conditions.
6   PAT testing – it is a requirement of the Electricity at Work Regulations that all portable electrical appliances are tested regularly, they should be marked with approved stickers and the inspection recorded in a log.

COSHH

The **Control of Substances Hazardous to Health** regulations require that assessments are made of all substances used in engineering. This assessment must state the hazards of using the materials and how to deal with accidents arising from misuse. Your wholesale supplier will provide you with this information as set out by the manufacturer in the form of either single sheets on individual substances, or a small booklet covering all the products in a range.

RIDDOR

The Reporting of Injuries, Diseases and Dangerous Occurrences Regulations 1995 require that certain information is reported to the Health and Safety Executive (HSE). This includes the following:

1   Death or major injury – if an employee or member of the public is killed or suffers major injury the HSE must be notified immediately by telephone.
2   Over-three-day injury – if as the result of an accident connected with work an employee is absent for more than three days an accident form must be sent to the HSE.
3   Disease – if a doctor notifies an employer that an employee suffers from a reportable work-related disease then this must be reported to the HSE.
4   Dangerous occurrence – if an explosion or other dangerous occurrence happens, this must be reported to the HSE, it does not need to involve a personal injury.

Maintain the Health, Safety and Security of the Work Environment

It is the duty of every employee and employer in the engineering industry to comply with the statutory regulations relating to health and safety and the associated guidelines which are issued by the various government offices. That means you must work in a safe and sensible manner. An engineer is expected to follow the health and safety recommendations of his/her employer; employers are expected to provide a safe working environment and advise on suitable safe working methods.

Health and Safety Law States that Organisations Must:

- provide a written health and safety policy (if they employ five or more people);
- assess risks to employees, customers, partners and any other people who could be affected by their activities;
- arrange for the effective planning, organisation, control, monitoring and review of preventive and protective measures;
- ensure they have access to competent health and safety advice;
- consult employees about their risks at work and current preventive and protective measures.

Tech note

Everybody in an organisation has a duty of care related to health and safety. The HSE may bring about prosecutions, or lesser prohibitions subject to timed actions – for instance being given a short period of time to rectify a machine fault. However, the final consequences can be devastating for a firm and its employees, possible outcomes are:

- Unlimited fine
- Imprisonment
- Closing down of the business
- Disqualification from working in that job or type of business.

You must work in a safe manner or you are breaking the HSWA and are liable to a fine, and possible imprisonment and maybe disqualification from working in that job. This means following the safe working practices which are normally used within the industry. The guidelines published by the Health and Safety Executive (HSE) and engineering textbooks usually identify industry-accepted safe working practices. Examples of important procedures are:

1   Always use axle stands when a vehicle or other heavy piece of equipment is jacked up.
2   Always use an exhaust extractor when running an engine in the workshop.

3   Always wear overalls, safety boots and any other personal protective equipment (PPE) when it is needed, for example safety goggles when grinding or drilling and a breathing mask when working in dusty conditions.

4   Always use the correct tools for the job.

When working on any system which contains fluids it is good practice to use a drip-tray to catch any possible spillages, this saves having to clean the floor as well as ensuring that all the used oils and fluids are disposed of safely, that is, you can pour them from your drip-tray into your disposal container. The Environment Protection Act requires that you dispose of used oils and other fluids in a way which will not cause pollution. In practice many companies use specialised companies to collect oils and other wastes on a regular basis.

Exhaust fumes are very dangerous, they can kill you. Small intakes of exhaust fumes will give you bad headaches, and over time can cause lung and/or brain diseases, so ensure that you do not run an engine in a workshop without an exhaust extractor. Also ensure that the extractor pipe is correctly connected and is not leaking.

The airline used in most workshops operates at between 100 and 150 psi (7 and 10 bar), this is a very high pressure, so it must be handled with great care. When you are using an airline always wear safety goggles to prevent dust entering your eyes. You must not use an airline for dusting off components, especially brake and clutch parts as the very fine dust can cause damage to the throat and lungs. Before you use an airline ensure that the coupling is fitted firmly into the socket and that the pipe is not leaking anywhere along its length. Any damage or leaks should be immediately reported to your supervisor or manager so that they can be repaired. The high pressure of the air can quickly turn a small leak in an airline into a large gash which in turn may make the airline whip around and cause damage to colleagues or a piece of equipment. Another area of potential danger is when using electrical equipment such as an electric drill, hand-lamp or grinder. Most mains-operated equipment runs at 240 volts; an electric shock from such a voltage is most likely to kill you straight away. Some companies use 110-volt equipment which is operated through a transformer, this is much safer, especially if the transformer is fitted with an overload cut-out. Hand-lamps should operate at 50 volts, or preferably 12 volts to give the highest level of safety. Plugs should only be fitted to electrical equipment by skilled persons, at the same time a fuse of the correct amperage rating should be fitted and the equipment tested and logged in accordance with the Portable Appliance Testing Regulations (PAT testing). PAT-tested equipment should be numbered and carry a test date label. Before you use any electrical equipment visually check it for signs of damage and check that the cable is not frayed or split. Then ensure that you plug it into the correct voltage outlet. Do not attempt to use any electrical equipment which you suspect may be faulty; report the fault immediately to your supervisor.

Where Identified Hazards Cannot be Removed, Appropriate Action is Taken Immediately to Minimise Risk to Own and Others' Health and Safety

This section is about those situations where the hazards cannot be readily removed, that is, how do you behave in accident situations, or when equipment malfunctions and you can see an accident about to happen? Most engineers are only likely to encounter such problems every few years, but the professional is the person who can save the day. The following is an example of where a service manager colleague came to the rescue. The central locking and the car alarm on a vehicle malfunctioned. A small child was trapped in the vehicle; it was a very hot day at a local car boot sale. The mother and child were hysterical, the father had gone for the fire brigade; other members of the public just watched. The mother shouted for help. My quick-witted friend grabbed a screwdriver off one of the stalls, inserted it behind the rear quarter-light rubber and levered out the glass, put his hand inside the car, opened the door and released the child. The panic was over.

There were other ways in which this situation could have been dealt with, but this one was acceptable because it provided a very quick solution and caused the minimum amount of damage to the vehicle. The important point is that people come first and property second, although the amount of damage to the property should be kept to the minimum.

Working by the roadside is always hazardous, but you can minimise the risks by following a few simple rules:

1   Always wear a high-visibility safety vest.
2   Use the warning triangle and hazard lights.
3   Use the flashing lamp on your mobile equipment.
4   Only work on hard and level ground.
5   Always use props when canopies or covers are raised.
6   Think through the possible hazards, never take risks.

Often it is better to do nothing than cause damage; this is referred to as preserving the situation. Many times things look different after a cup of tea, or you have had time to check it out with a colleague.

Dangerous Situations are Reported Immediately and Accurately to Authorised Persons

As a trainee in the engineering industry your company will require you to report any dangerous situations to your supervisor; this will be a person that you know as the charge-hand, foreperson or service manager. Any internal matter should in the first instance be reported to one of these people – you will know who this is from your induction training. However, if you are working alone or the matter is not a company one, then you must inform the relevant authority. The four emergency services in the UK are Police, Fire, Ambulance and Coast Guard. To call them use any telephone and dial 999.

Suppliers' and Manufacturers' Instructions Relating to Safety and Safe Use of All Equipment are Followed

Many pieces of engineering equipment are marked 'only to be used by authorised personnel'. This is mainly because incorrect use can cause damage to the equipment, the work-piece or the operator. Do not operate equipment which you have not been properly trained to use and have not been given specific permission to use.

The suppliers of engineering equipment issue operating instructions and as part of your training you must read these instruction booklets so that you will understand the job better. You will also find that certain safety instructions are marked on the equipment. The hoist or hydraulic jack and other lifting equipment are marked with the Safe Working Load (SWL) in either tonnes or kilograms. You must ensure that you do not exceed these maximum load figures. Some items of equipment have two-handed controls or dead-man grips – do not attempt to operate these items incorrectly.

Approved/Safe Methods and Techniques are Used when Lifting and Handling

Do not attempt to manually carry a load which you cannot easily lift and which you cannot see above and around. The maximum weight of load that you should lift is what you feel comfortable with, this is usually around 20 kilograms, but as a trainee this may still be too heavy for you.

When you are lifting items from the floor always keep your back straight and bend your knees. Bending your back whilst lifting can cause back injury. If you keep your feet slightly apart this will improve your balance. It is always a good idea to wear safety gloves when manually lifting.

Hoists and jacks are available for lifting heavy items; for moving equipment and heavy components you should have either a trolley or forklift.

Table 2.1 PPE Usage

No	PPE	Usage
1	Cotton overalls	All the time
2	Safety footwear	All the time
3	Disposable gloves	Dealing with dirty or oily items
4	'Rubber gloves'	Operating the cleaning bath
5	Reinforced safety gloves	Handling heavy, sharp edged items
6	Dust mask	Rubbing down
7	Breathing apparatus and paper coveralls	Spray painting or using other materials
8	Goggles	Using grinder, drill or other machine tools
9	Waterproof overalls and boots	Steam or pressure washing

You are advised to seek the assistance of a colleague when moving a heavy load, even when you are using lifting equipment.

Required Personal Protective Clothing and Equipment are Worn for Designated Activities and in Designated Areas

Table 2.1 lists typical items of personal protective equipment (PPE) and states when they must be worn.

You will often see safety notices requiring you to wear certain PPE in some areas at all times, this is because other people are working in the area and you may be at risk. Hard hats are sometimes required when working in some areas.

Injuries Involving Individuals are Reported Immediately to Competent First Aiders and/or Appropriate Authorised Persons and Appropriate Interim Support is Organised to Minimise Further Injury

Should there be an accident the first thing to do is call for help. Either contact your supervisor or a known first aid person. Should any of these not be available, and it is felt appropriate, call for your local doctor or an ambulance.

You are not expected to be a first aid expert, nor are you advised to attempt to give first aid unless you are properly qualified. However, as a professional in the engineering industry you should be able to preserve the scene, that is, prevent further injury and make the injured person comfortable. The following points are suggested as ones worth remembering:

1 Switch off any machinery or power source.
2 Do not move the person if injury to the back or neck is suspected.
3 In the case of electric shock turn off the electricity supply.
4 In the case of a gas leak, turn off the gas supply.
5 Do not give the person any drink or food, especially alcohol, in case surgery is needed.
6 Keep the person warm with a blanket or coat.
7 If a wound is bleeding heavily, apply pressure to the wound with a clean bandage to reduce the loss of blood.
8 If a limb has been trapped, use a safe jack to free the limb.

Visitors are Alerted to Potential Hazards

The best policy is not to let customers into the workshop – many engineering companies have a notice to this effect on the workshop door.

It is always essential to accompany visitors when they are in the workshop, this way you can advise them in the event that they may do something potentially dangerous or if there is a hazard of which they may not be readily aware.

Injuries Resulting from Accidents or Emergencies are Reported Immediately to a Competent First Aider or Appropriate Authority

If a person is injured the first action must be to ensure that first aid is given by a competent first aider or other suitable person. Most companies have a designated first aider who is trained to deal with accidents and emergencies. If your company has no such person on the staff then you will have a designated person who you must contact in the event of a colleague being injured. That person may be your supervisor or another senior member of the staff. If no manager or other senior person is available you should either dial 999 for an ambulance or telephone your company doctor, then inform the workshop manager.

Incidents and Accidents are Reported in an Accident Book

By law all companies are required to maintain records of accidents which take place at work. These records are usually kept in an accident book. Accident books may be inspected by HSE inspectors; they must be kept for a period of at least three years from the date of the last entry.

The information which is required to be recorded in the accident book is:

- Name and address of injured person
- Date, time and place of accident/dangerous occurrence

- Name of person making the report and date of entry
- Brief account of accident and details of any equipment/substances which were involved.

It is always a good idea to keep a notepad to help remind you which way round things go when working on unfamiliar equipment, this would also be useful for making any other notes, such as those about an accident.

Where There is a Conflict Over Limitation of Damage Priority is Always Given to the Person's Safety

You can always buy a new wing for a car, but you cannot buy a new arm for a mechanic. In the event of an accident people come first. For instance, if a building is on fire, do not re-enter to retrieve your belongings, wait until the fire is out and there is no risk before going back into the building. If a car is about to fall off a jack, get out of the way, do not try to catch the car with your hands or some such other dangerous action.

Professional Emergency Services are Summoned Immediately by Authorised Persons in the Event of a Fire/Disaster

An authorised person is somebody who has the task of carrying out a specific job. Anybody may call the emergency services if they are needed.

The Four Emergency Services are:

1. Police
2. Fire
3. Ambulance
4. Coastguard

All are called by dialling 999 on an outside-line telephone. The emergency services operator will ask you which service you require. In certain cases the police will automatically be called, for instance in the case of a severe fire.

All emergency telephone calls are recorded on tape at the telephone exchange. You will be asked for your name, the place where the emergency is and where you are calling from. With the introduction of electronic telephone exchanges the number which you are calling from is automatically recorded, and you will be asked for the number to help confirm that your call is not a hoax.

Many companies have a direct telephone line to the fire station, and these automatically call the fire service if a fire is detected by sensors or by breaking the glass of a fire alarm. In such cases, if the fire service is called out and there is not a fire, they may charge the company a large fee. So, do not tamper with such a device unless you are authorised to do so or there is a dangerous fire.

Alarm/Alert/Evacuation Systems are Activated Immediately by Authorised Persons

Generally, only senior staff (managers) are allowed to operate alarms and other forms of alert/evacuation systems. This is because of the costs which may be involved if the fire service is called out wrongly and the damage which may occur if staff and/or customers panic.

In the case of a fire the normal alarm is a form of siren or bell. For other emergencies, say a serious injury, an audible warning from a speaker announcement system may be used. In a workshop these are usually operated from the office.

In the Event of Warnings, Procedures for Isolating Machines and Evacuating Premises are Followed

If you hear a fire/emergency warning you must follow the company's evacuation procedure – this is usually stated on the workshop wall. If you hear a fire alarm a typical evacuation procedure is:

1 Shut off the electricity by pressing the emergency stop button.
2 Leave the premises by the nearest route, go to assembly point, or muster point, which is in the customers' car park.
3 Do NOT re-enter the building until your supervisor tells you that it is safe to do so.

In the event of discovering a fire, raise the alarm by breaking the glass of the alarm button.

Reports/Records are Available to Authorised Persons and are Complete and Accurate

The Social Security (Claims and Payments) Regulations 1979 require employers to maintain an accident book as well as regulation 7 of the HSWA. This book requires brief details of any accident or dangerous occurrence to be recorded. An approved book BI 510 is available from the HSE direct or through most good book stores. For more detailed information, HSE Form 2508 should be completed. HSE staff have a statutory right to see a completed accident book or Form 2508, and they may also ask for further information. If you are personally involved in an accident you are advised to keep a copy of the book entry and any completed forms as well as your own notes on the event. These may be useful in the event of legal proceedings.

Machines and Equipment are Isolated, Where Appropriate, from the Mains Prior to Cleaning and Routine Maintenance Operations

You must always isolate an electrical machine from the mains supply before either cleaning it or carrying out any maintenance or repairs. There are two reasons for this: first, if you touch an electrically live part you may get an electric shock; second, the machine may be accidentally started which could cause injury or damage.

With portable electrical appliances this simply means switching off and taking the plug out of the socket.

With fixed machinery, for instance a pillar drill, you will need to switch off the power --supply at the isolator switch. This is usually found on the wall near the machine. Isolating this way is fine while cleaning the machine, but for carrying out maintenance or repair work it is advisable to remove the supply fuse from the isolator box. With the fuse removed the machine cannot be restarted if the isolator is accidentally turned on by a colleague who confuses the isolator for the one on an adjacent machine.

Safe and Approved Methods for Cleaning Machines/ Equipment are Used

There are three main items of cleaning equipment used in the workshop: the cleaning bath (or tank), the pressure washer and the steam cleaner.

The cleaning bath uses a chemical solvent, this is usually used for cleaning dirty/oily components. The components are submerged in the solvent and dirt is loosened with a stiff-bristled brush.

The pressure washer is used for cleaning the mud and grime off equipment; water at very high pressure will clean off mud. For hard-to-remove dirt, detergent can be added to the pressure washer. The steam cleaner, often referred to as a steam jenny (jenny = generator), produces hot pressurised water with the option of detergent. This is used for removing very stubborn grease and dirt, like that found on the underside of high-mileage goods vehicles and old equipment.

When cleaning portable electrical appliances be careful not to get water or other liquid on the plug, this could cause a short circuit.

The mechanical parts of fixed machines may be cleaned with solvents, then dried with absorbent paper towel.

Appropriate Cleaning and Sanitising Agents are Used According to Manufacturer's Instructions

Before using any solvent, detergent or sanitising agent such as bleach you must read both the label on the container and the COSHH sheet which the manufacturer or your company has prepared.

Solvent should only be used in the cleaning bath for which it is designed.

The pressure washer or steam jenny should only be used with the recommended detergent.

Electrical items can be cleaned with one of the many aerosol sprays which are available for this purpose, but the volatile fumes which are given off must not be breathed in.

You should remember that all cleaning agents should be kept away from your mouth and eyes, and contact with your skin may cause irritation or a more serious skin disease. Always wash your hands and any other exposed areas of skin with toilet soap after carrying out a cleaning task.

Used Agents are Safely Disposed of According to Local and Statutory Regulations

The Environmental Protection Act (EPA) and local by-laws in most areas require that used cleaning solvents must be disposed of safely. This means that they must be put into drums and either collected by a refuse disposal firm or taken to a local authority amenity site where they are put into a large tank for bulk incineration. Several local authorities, for instance Surrey and Hampshire, are now looking at ways of using the energy produced by burning waste material to produce electricity. Emptying used solvents into the drain can lead to a heavy fine or even imprisonment.

Detergents are by their nature biodegradable, that is, they break down, do not build up sludge and will not explode, unlike solvents. However, if you use large quantities of detergents, wash bays which are fitted with the correct type of drainage system should be used.

Machinery, Equipment and Work Areas are Cleaned According to Locally Agreed Schedules

In your company's Health and Safety Policy document there will be reference to the cleaning of the floors and equipment in the workshop and general amenities such as toilets and rest areas. Also there will be maintenance and repair records for the workshop equipment which will include a regular schedule of cleaning and inspection.

Most companies work on the basis of sweeping down fixed machinery and floors at the end of each day, unless the generation of dirt requires more frequent attention.

On a weekly basis there will be a more thorough cleaning programme which may include window cleaning and wet-cleaning certain areas.

Workshop equipment is usually cleaned and inspected on a monthly basis unless there is reason, such as a fault, for a more regular treatment.

Appropriate Safety Clothing and Equipment are Used When Working with Hazardous Cleaning Agents and Equipment

To protect yourself from the cleaning agents which you are using you must, where appropriate, wear personal protective equipment (PPE). Most cleaning agents are poisonous and cause irritation or more serious complaints if allowed to come into contact with your eyes or skin.

Whenever you are working in a workshop it is expected that you wear cotton overalls and safety footwear. In addition, the HSWA requires that employers provide and employees wear the appropriate PPE for hazardous jobs such as using cleaning equipment. The general requirements are as follows:

1 Cleaning bath – rubber protective gloves which extend over the user's wrists, goggles and plastic apron. Avoid getting solvent on your overalls as this can lead to skin irritation, be especially careful not to put solvent-soaked or oily rags in your overall pockets.
2 Pressure washer –protective rubber gloves and goggles, waterproof (plastic) over-trousers and jacket, and finally rubber boots (wellingtons). The idea is to be able to take the waterproof gear off and be dry underneath.
3 Steam cleaning plant – the hazard here is that as well as being wet the water is scolding hot. So the waterproof clothes must be of such a manufacture that they will protect the wearer from the high-temperature, high-pressure steam. This means thick and strong over-trousers, coat, boots, gloves and a hat. A full-face mask is used to give complete protection.

Skills and Questions

Care for Health and Safety, and the environment is something which MUST be part of EVERY job that you do EVERY day. However, these skills need developing, and do not always come easily – keep practising until Health and Safety are second nature for you. Meanwhile try these questions:

1   State five basic rules concerning dress and behaviour which demonstrate personal safety in the workshop environment.

2   List five necessary precautions for safety in the workshop and describe each one briefly.

3   What is meant by a skin care system as used in the workshop?

4   Explain the importance of eye and face protection in the workshop environment.

5   Explain the importance of protective clothing for an engineer.

6   Explain the significance of headwear and footwear while working in the workshop.

7   State the minimum noise level at which ear protection must be used.

8   With the aid of a diagram, explain the fire triangle.

9   Identify the correct colour code for the following fire extinguishers: water, foam, CO_2 and dry powder.

10   Explain the precautions which must be taken when handling toxic substances in a workshop environment.

Chapter 3

Design Processes and Practices

Engineering Drawings and Graphical Language

Engineering drawing is a means of communication without language problems, the drawing describes the object and its features. Added to the basic drawing are a number of symbols and letter and/or numerical abbreviations, or terms. The symbols and abbreviations, along with the drawing conventions are part of a set of International Drawing Standards. When these are used the drawing can be understood by any engineer in any country. This overcomes language barriers and transmits the information clearly and in detail.

Sketching

Sometimes it is necessary just to explain a simple point, or describe a situation, often this can be done by means of a simple sketch. This is particularly useful where two engineers need to communicate at a distance. One can make a simple sketch, take a photograph of it using the mobile telephone camera and message it to the other. Sketching is a very useful skill to learn, and it requires very little in the way of equipment.

Drawing Equipment

Lots of different equipment is available, at different levels of quality and price points. The following are suggested as a starting point for students:

- Pencils: 4H, 2H, HB, 2B, 4B
- Pencil sharpener
- Eraser
- 30 cm (12 inch) ruler
- 30°/60° set square
- 45° set square
- Compass

DOI: 10.1201/9781003284833-3

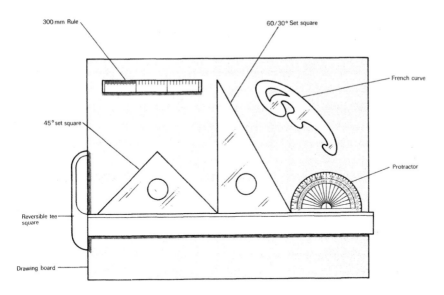

Figure 3.1 Drawing board and instruments.

Tech note

Pencils are classified according to their hardness – H, the higher the number the harder the pencil, and their blackness – B – the higher the number the blacker will be the line.

A drawing board will be required, there are lots of options in these in regards of size and use of parallel motions or 'T' square. A3 is the most common and useful size. Paper sizes in common use go from A0 to A8, the smaller the number the larger the sheet. Each sheet size doubles on area by doubling the length of the shorter size.

Paper size	Millimetres
A0	841 × 118
A1	594 × 841
A2	420 × 594
A3	297 × 420
A4	210 × 297
A5	148 × 210

Paper size	Millimetres
A6	105 × 148
A7	74 × 105
A8	52 × 74

The 4H and 2H pencils are used for drawing guide lines and dimension lines. Guide lines should be drawn very lightly, such that they are visible when you look closely, and be invisible, unseen, when you lean back in your chair. HB is for normal outlines of the components which you are drawing, and 2B and 4B where you need emphasis on a drawing. The full range of pencils runs from 9H to 9B, if you wish to make more detailed illustrative sketches.

Types of Drawings

Isometric – means of equal measure – (iso – equal; metric – measure) – with this type of drawing the component, or part, can be seen on three faces, these are front, side and top. This gives a good indication of the bulk of the object.

Start with a point on your base line, the two sides are drawn at 30° to the base line – use a set square – so that the included angle is 120° between the two lines. All receding – going backwards, or going away – lines are at the same angle on each side. The vertical lines are at a 90° angle to the base line. This is sometimes called a right angle (rt Δ), or 'normal to', or perpendicular.

Oblique – is looking at an item from an angle. This angle is neither parallel to, nor at 90° to, the face. Start by drawing the face straight on. Then draw the receding face at 45° – chosen as it is midway between parallel and 90° – in other words oblique.

Cabinet and cavalier oblique projection – drawings to represent furniture became popular in sales catalogues in the 18th century. When drawn in the original way, the furniture looked squat, or fat. The proportions did not look correct. That type of drawing is what is now called cavalier projection. The cabinet makers discovered that if they shortened the receding lines, the ones going back at 45°, their cabinets looked like they should. Hence the name cabinet projection.

Tech note

The concept of cavalier and cabinet projection differences is best seen by sketching. On the same base line draw two 25 mm squares about 50 mm apart. Now convert them both into cubes by drawing receding lines. One full length, 25 mm, the other half length at 12.5 mm. You should see the difference clearly – you might want to get your ruler out to check.

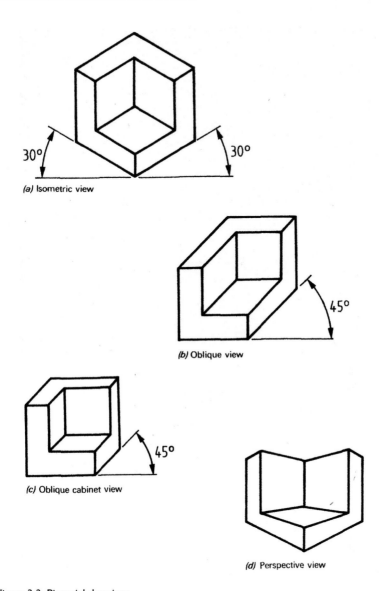

(a) Isometric view

30°   30°

(b) Oblique view

45°

(c) Oblique cabinet view

45°

(d) Perspective view

Figure 3.2 Pictorial drawings.

Orthographic projection – ortho means correct, graphic is making a mark. If you are making, or repairing, an object you will probably need more information than an isometric or oblique projection can provide. You might want to know all about the other three faces.

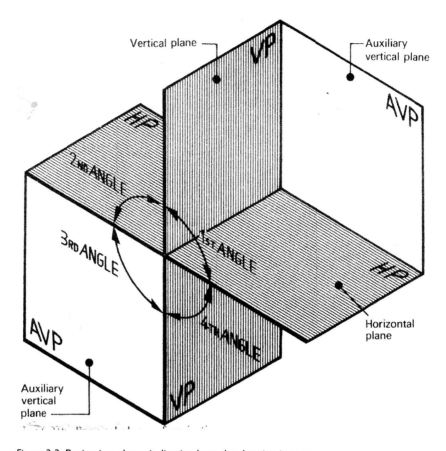

Figure 3.3 Projection planes indicating how the drawing is seen.

Tech note

Most objects have six faces: top, bottom, two sides, front and back.

An orthographic projection drawing is used, as its name suggests, for accuracy, correctness. An orthographic drawing usually shows three sides of an object, in some cases all six sides are drawn. There are four different ways of conventional representations possible, although only two are in common use, these are called first angle and third angle.

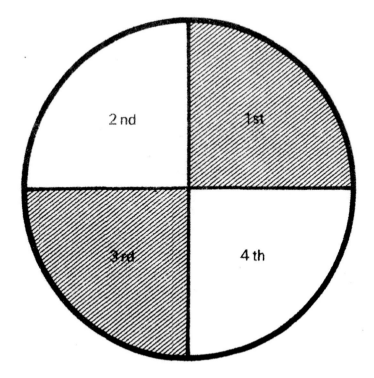

Figure 3.4 Showing drawing quadrants.

Tech note

In engineering drawing offices most products are drawn in 3D, so they can be flipped around – that means drawing all the six sides together.

Standards – to ensure that drawings are readily understood, and that the different marks on them mean the same thing to different engineers, that is, so that the engineer can interpret them accurately, drawing offices follow a set of standards. These are conventions in drawing. In the UK the standard is BS8888, this is used worldwide by manufacturers. The European equivalent is ISO128. Both standards are very similar. BS8888 is a follow-on from BS308 which was the first drawing standard developed in the world in 1927. These standards are available online as PDF documents.

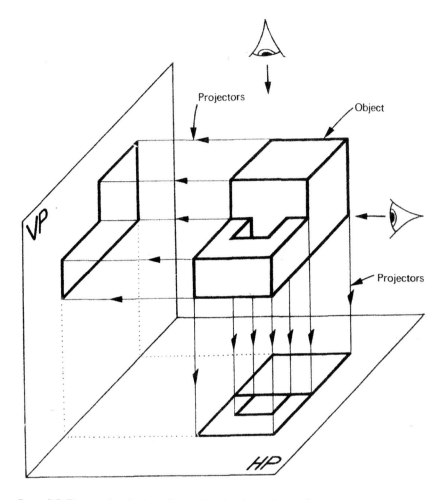

Figure 3.5 First angle viewing planes showing how the surfaces are projected on to the paper.

BS8888 covers:

- Scale
- Dimensioning
- Terms and definitions
- Tolerances
- Geometric tolerancing
- Indications of surface texture
- Lines and arrows

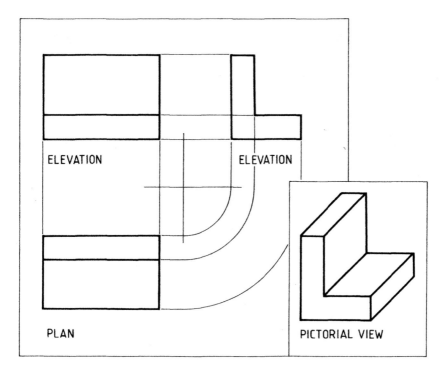

Figure 3.6 An example of a 1st angle drawing.

- Lettering
- Projections
- Views
- Symbols
- Abbreviations
- Representation of features and components

Tech note

BS is British Standard; BSI is British Standards Institute who write standards; ISO is International Standards Organisation; EN is European Norm. ANSI is the American National Standards Institute. All standards are very similar.

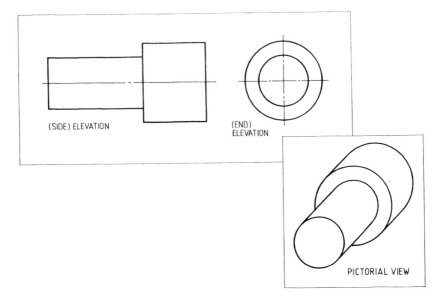

(SIDE) ELEVATION

(END) ELEVATION

PICTORIAL VIEW

Figure 3.7 An example of a 1st angle drawing with two sides only.

Dimensions – the title box is usually used to indicate the units for the dimensions, typically in millimetres (mm) in the UK, Europe and Asia. In the USA this may be inches or millimetres. In the USA, using inches, the fractions are usually given in tenths of an inch. Be aware that where drawings are less than full-size, either to scale, or just illustrative, the dimensions may be in a variety of units.

Scale – small items on paper drawings may be full size, others may be to scale. You should always check the scale and check the units. A popular scale for model making is 'O' gauge. This is 7 mm to a foot. To make this easy to calculate scale rulers can be obtained, these are marked in the actual size.

Drawing Terminology

- Angle – angle in degrees to an indicated section.
- Assembly drawing – a drawing which shows the positions of individual items used to form a complete assembly.
- Detail drawing – often larger scale, bigger in size, than the rest of the drawing, to illustrate particular points, such as how it fits other parts.
- Elevation – term used to describe parts of an orthographic projection: side elevation, end elevation, plan view.

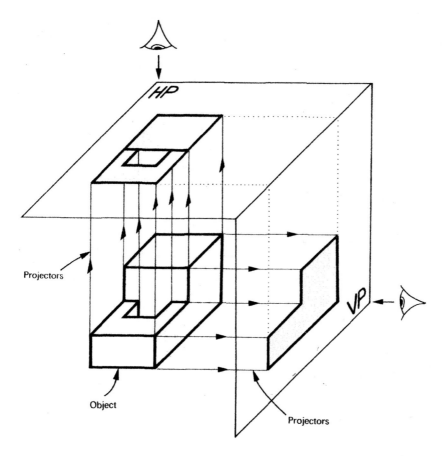

Figure 3.8 Third angle viewing planes showing how the surfaces are projected on to the paper.

- PCD – pitch circle diameter, where studs or holes fit on a round item – such as car wheel studs.
- Production drawing – drawing with sufficient detail for it to be made, also called a manufacturing drawing.
- Radius – radius of a curve, or hole shown on a drawing with the letter 'r'.
- Sectional drawing/view – usually used to show inner detail.
- Sketch – hand drawn to illustrate a point, usually not to scale, and not detailed.
- Surface finish – the required smoothness, or otherwise, of a component.
- Tolerance – the minimum and maximum differences from a given face or datum.
- View – looking at a component from one position, see elevation.

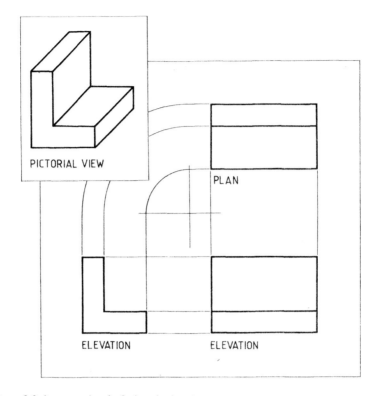

PICTORIAL VIEW

PLAN

ELEVATION

ELEVATION

Figure 3.9 An example of a 3rd angle drawing.

Limits, Fits and Tolerances

The dimensions on the drawing may also indicate the maximum and minimum sizes allowed to ensure easy assembly.

Surface finish – or roughness values, these may be indicated in values of micrometres – μm which is 0.000001 m.

Geometric Tolerances

Even though a machined item may be within tolerance for size, it may not be geometrically correct. The six points of geometric tolerancing are:

1. Flatness – to machine a perfectly flat surface is both expensive and time consuming, so it is usual to indicate the level of flatness needed. This is very important on petrol and diesel engines where the cylinder head fits on to the cylinder block.

2. Concentricity – when two circles, or cylinders, are machined they will need to have some level of concentricity. A good example of this is the fitting cylinder liners in car and truck engines.

3. Circularity – a piston fitting in a cylinder bore needs to be absolutely circular to prevent the by-passing of gases.

4. Cylindricity – the cylinder in an engine needs to be smooth all the way down.

5. Perpendicularity – some surfaces will need to be at exactly 90° to each other. An example is tall buildings, if they are not perpendicular they will not be stable.

6. Straightness – most engineered components require a high level of straightness, for example bicycle frame tubes.

Abbreviations used on drawings

Across flats	A/F
Centres	CRS
Centre line	CL
Chamfered	CHAM
Cheese head	CH HD
Countersunk	CSK
Countersunk head	CSK HD
Counterbore	C'BORE
Diameter	DIA or Ø
Drawing	DRG
Figure	FIG
Hexagon	HEX
Hexagon head	HEX HD
Material	MATL
Number	NO.
Pitch circle diameter	PCD
Radius	R
Screwed	SCR
Specification	SPEC
Spherical diameter of radius	SPHERE R or Ø
Spotface	S'FACE
Standard	STD
Undercut	U'CUT

Electrical and electronics – drawings in these areas are mainly centred around circuit diagrams, these should be drawn using the appropriate symbols and conventions.

1 Bevel drive housing
2 Pivot stub
3 Pivot bearing
4 Pivot shaft
5 Locknut
6 Plastic cap
7 Clamp
8 Gaiter
9 Clamp
10 Torque arm
11 Bolt
12 Bush
13 Washer
14 Nut
15 Circlip
16 Washer
17 Front torque arm mounting
   bolt/brake pedal pivot
18 Bush
19 Washer
20 Nut
21 Rear suspension unit
22 Bolt
23 Washer
24 Bush
25 Nut
26 Washer
27 Bush
28 Swinging arm
29 Plastic cap
30 Locknut
31 Pivot shaft
32 Pivot bearing
33 Grease retainer - where fitted
34 Clamp
35 Gaiter
36 Clamp

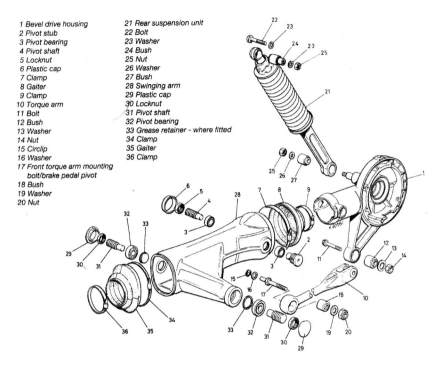

Figure 3.10 An example of an assembly drawing.

Plumbing – offers the opportunity to draw actual items as well as diagrams for the pipework connections.

Schematic and assembly drawings – these are specialised drawings, often in the form of pictorial sketches, they usually do not follow any conventions; but aim to get a message over, how to assemble, or fit something without using any words.

CAD/CAM – computer-aided design and computer-aided manufacturing is used for most engineered products. This gives greater speed, accuracy and economy. This topic is not covered in this book.

Design Process

Most engineering companies have a design office or studio. Large companies may have large design facilities separate from manufacturing and sales. The design process may be started in many different ways, it is closely related to project management, in fact, most design work is carried out as a project.

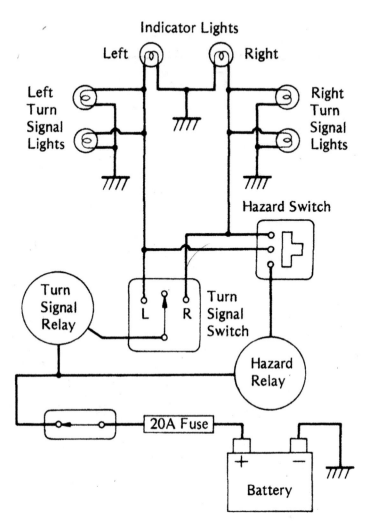

Figure 3.11 An example of a wiring diagram.

Large companies may have several design teams working on the same initial concept, or product development. The reason for this is to get multiple solutions to ideas, generally more people means more creative solutions.

The two most recognised forms of the design process are **market pull** and **technology push.**

Market pull is where the market research team, working with customers in focus groups, and long-term testing identify customer needs and wants. This may relate to making a completely new product to satisfy a perceived need, or updating an existing product to rectify a fault, meet a specific need, add additional functionality, or meet a specific aesthetic – make it appealing to particular groups of people. The design team will then look at a variety of ways to meet the needs identified by the market research team. They will, at the same time, apply costings and develop a prototype. The market researchers will re-visit the focus groups and obtain further feedback. They may also have a short production run to gauge market appetite – in other words find out if they sell.

Tech note

A focus group is usually a group of people, it may be known existing customers, or ones chosen because they meet the socio-economic specifications of people who are likely to purchase the product that is being designed.

As an example, in the automotive industry, cars are often made in a variety of specifications, the specifications will alter the costing involved and so the sales price point. The top of the range model may cost over double the price of the base model. Which sells most will inform manufacturing choices. Also to be borne in mind is the profit made by each model and the availability of the necessary components.

Technology push is where a company carries out research and develops product which they think that customers will want, and buy, in the future. They look for customers who are early adaptors, people who want to try new things. Such products on initial entry to the market are usually at a premium price. The high price usually is to cover the costs of the initial product research and design.

The design process may take many forms, usually they follow a pattern like this in a linear format, passing from stage to stage:

1. Market needs/competition from other manufacturers/technology push/ observed flow, or need to develop existing products.
2. Analysis of situation or problem and development of a new design specification.
3. Development of a new design concept, the principles of how it addresses the identified market needs. This may result in a model, or maquette, being built for aesthetic visualisation. This will include general layout and maybe 3D modelling.
4. Design detail and specification producing a pre-production model. This model may be used by the marketing team and selected focus groups.
5. Final design for production and sales.

After this, the process will begin again for the next generation of the product.

The design process, no matter whether it is market pull – that is customer led – or technology push – that is a perceived concept – will start with some form of **product design specification.**

The product design specification is a comprehensive document describing exactly what is needed in detail. Typical headings in the product design specification are:

- Working title, or name of product, usually with a code name that can identify it from previous models, or competitors. This is usually a mixture of letters and numbers which are easily remembered. An example is the Ferrari 812GTS, the 8 refers to the 800 HP power output, the 12 is the number of engine cylinders, GT means that it is a grand touring car – seats and luggage space, S stands for spider – an open car.
- Functionality of the product – what it can be used for, multi-functionality is often desirable.
- Performance – this is often a balance of speed and power against weight and cost.
- Quality – in terms of aesthetics and reliability.
- Health and Safety issues for both manufacturing and usage.
- Conformability with rules and regulations in the intended market.
- Price point – compared to similar products in the market.
- Materials – this relates to strength, aesthetics and costs.
- Production methodology.

Tech note

Many of the points relating to product design are discussed in other chapters of this book, the reader is directed to the index to identify page numbers.

Skills and Questions

A common cry from students is, 'I can't draw.' Just like using any other tool you need to learn how, and then practise. When you start to get the skill, you will find it very useful for all your lessons, sketching an engineering component is usually much easier that describing it in words.

1. To get started, take a photograph of any engineered object, print it out A4 size, then either using tracing paper, or a light box and plain paper, trace out the object. This will help you to see the lines which make up the object.

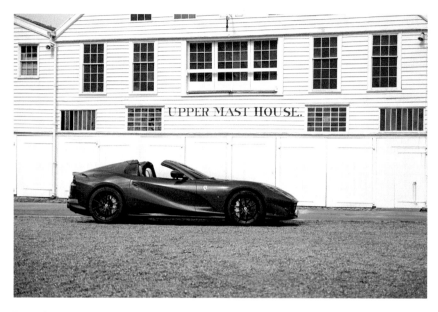

Figure 3.12 Ferrari **812GTS**.

2. Using a ruler and pencil sketch a square, or oblong, object in both iso-metric and oblique form from your measurements – use both cavalier and cabinet projections.
3. Draw any engineered object in either 1st or 3rd angle orthographic pro-jection. Make sure that you add a detailed title box.
4. Set yourself up an engineer's notebook, throughout your course pro-gramme make sketches of items which interest you and things that you make, or repair.
5. Take photographs of engineering items, tools, equipment and other items, add sketches and labels to them throughout your course, save these in your notebook.

If your career goal is in plumbing, electronics, construction or another field of engineering then tailor your sketches and your notebook to those areas.

Chapter 4

Maintenance, Installation and Repair Practices

This area of engineering is one of the biggest and highest paid areas of employment in the world. There are a number of different names for this area of work, these include **Industrial Engineering** and **Operations Engineering**.

Tech note

The job titles include aircraft technician, computer repair engineer, cycle mechanic, hospital engineer, machine maintenance technician, maintenance engineer, marine engineer, motor mechanic, motorcycle mechanic, photocopy technician, race car mechanic, radio engineer, railway engineer, TV/video engineer, and many more. The skills and procedures within MIR are largely interchangeable, and it is very common for engineers to move between jobs in this area of engineering.

Job Roles

In all areas of MIR there is a requirement for a set of skills that apply to most engineers, this includes, not in any particular order:

- Organisational skills in terms of time keeping and having the correct tools and equipment ready when needed.
- Working in a clean and tidy manner.
- Reading and following instructions.
- Appropriate numerical skills related to common physical quantities such as size, weight, pressure and temperature.
- Correctly and safely use a range of tools and equipment.
- Able to both act upon, and make risk assessments.
- Keeping both accurate and timely contemporaneous records.
- Understanding the complete functionality of the engineering system which is being worked upon.

DOI: 10.1201/9781003284833-4

Figure 4.1 Deep socket.

Job Description for an Installation Engineer

Engineers who work on installing various machinery or equipment are called installation engineers. They are also able to manage smaller machinery and tools. Installation engineers must be constantly aware of Health and Safety regulations. Their jobs must not affect the safety of their clients, nor themselves.

Figure 4.2 Socket adapters.

Figure 4.3 Combination spanner set.

An installation engineer usually performs most of the following tasks:

- **Assessing risks,** this may involve PUWER 1998, LOLER 1998, MHO 1992, PPE 1992, EAWR 1989, CSR 1997 and many others.
- **Liaising with clients,** a good telephone manner is needed along with good personal presentation and level of hygiene.
- **Filling out job sheets,** tidy, detailed and accurate presentation is needed.

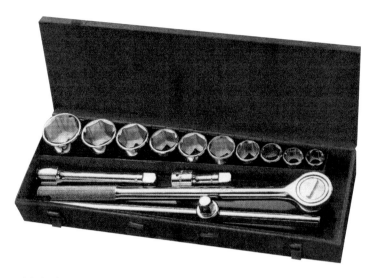

Figure 4.4 Socket set.

- **Installing equipment**, detailed knowledge of both the equipment and location is needed.
- **Reporting to manager**, good verbal skills are needed.
- **Adhering to regulations**, most equipment is now covered by rules and regulations, it is the professional responsibility of the engineer to ensure that these requirements are fully met and understood by the customer too.
- **Maintaining equipment**, being able to carry out maintenance procedures and explain them to others as needed.

Job Description for a Maintenance Engineer

The duties and responsibilities of a maintenance engineer vary depending on the industry in which their employer operates, but their main duties usually include:

- **Creating a maintenance plan** for the organisation's equipment, this may be time based, or dependent on other criteria such as the number of working cycles completed.
- **Establishing maintenance policies and procedures** – these will include Health and Safety and budgetary control.
- **Ensuring that maintenance technicians carry out planned maintenance** – this will include service interval documentation audits and checks.

Figure 4.5 Stud removal tool.

- **Conducting frequent quality checks** on equipment to ensure that performance is up to the required standard and prevent breakdowns.
- **Completing emergency maintenance** when equipment breaks down.
- **Managing the budget** for the maintenance of equipment and its replacement.
- **Keeping a record** of preventive and emergency maintenance carried out.
- **Ensuring the organisation complies with Health and Safety regulations** while maintenance takes place and at all times when it is being operated.

Breakdowns and Repairs

Both installation and maintenance engineers deal with breakdowns and repairs; but some breakdowns need specialist services. Examples of these are: roadside breakdowns of vehicles, it is mainly unsafe to try to carryout roadside repairs, so recovery vehicles are used to take the broken down, or accident-damaged vehicle, to a garage or accident repair centre. The AA and the RAC have staff specially trained in safe recovery techniques. Failure of an electricity power supply is dealt with by installing a temporary generator. The fire service deals with a wide range of emergencies, related to breakdowns and accidents. At sea, the Life Boat Service – in the UK this is the RNLI – and the Coast Guard Service are the recovery specialists. For large-scale emergencies the armed forces may be called in – the Royal

Figure 4.6 Aircraft spanners.

Engineers are trained in almost all aspects of engineering. All the emergency services have some skilled engineering staff.

Work Experience

The best way to find out about any job role is work experience; to give you a taster here are two written examples of the sorts of things you might do:

Racing Car Engineer

Working on race cars, inspect, test and rebuild

Safety first – When approaching a vehicle for the first time, especially a damaged motorsport vehicle, you must carry out a risk assessment. That is a mental assessment of the situation. If you are on duty at an event, maybe as a team mechanic or a marshal, you will be first on the scene. In this case you need to consider the following:

- Is it safe to get near the vehicle – think about location, other traffic, and other people – you must put your safety as first priority.
- The next priorities are making the scene safe, calling for help, and the application of first aid and perhaps the paramedics.

If you are approaching the vehicle after its recovery, even when it is in the pits or workshop, a mental risk assessment is needed.

Some typical examples of the dangers are:

The rally mechanic who opened the bonnet of a vehicle at the end of a forest stage to be scalded by the coolant from a radiator fracture – the coolant was at about 130°C. The NASCAR mechanic who decided to touch

Table 4.1 Using your Senses

Sense	Checks	Cautions
Sight	How does this vehicle look? Is it level and square? Are there any leaks or stains? Signs of damage or misuse?	Use eye protection
Sound	How does this vehicle sound? Listen to the different systems or parts	Use ear protectors
Smell	Is there a smell which might indicate a leakage or overheating?	Wear a mask
Taste		Not advised
Touch	Use your fingertips to check for damage or wear. Use a nail to check whether a blemish is raised or sunken	Use hand protection
Kinaesthetic	Feel the operation of controls or mechanical linkages for smoothness	Be prepared for the unexpected

the brake disc when the car came in the pits for a wheel change; he burnt his fingers. The discs were probably at about 800°C. Remember that you need to wear gloves to even touch the tyres.

Use your Senses

When inspecting a vehicle it is always good to use your senses, but do it with care.

Respect for Vehicles

As a technician you are responsible for the vehicle that you are inspecting. Therefore you must not cause damage to the vehicle, even if the vehicle that you are inspecting is seriously damaged. You never know what repairs or salvage may take place. You are expected at all times to:

- Use seat covers
- Use floor mat protectors
- Use wing covers
- Protect from bad weather
- Jack up and support the vehicle safely using appropriate jacking points
- Ensure that the systems are treated with care
- Remove finger marks.

The following tables are to give you a practical guide to inspecting motor-sport vehicles – you may choose to use them as a guide and to make up your own checklists. Most teams have some form of checklist – this is likely to be computer based.

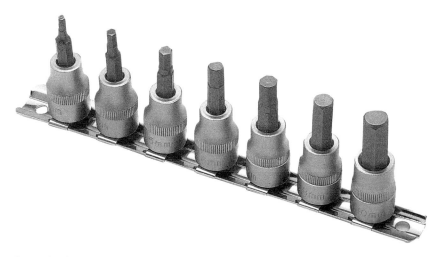

Figure 4.7 Allen key drives.

Figure 4.8 Bearing puller set.

Race Car Preparation Workshop Equipment

The sorts of tools and equipment that you will find in the workshop.

Bench work area – the bench work area is often around the out-side of the vehicle bays, using metal-topped benches with drawers and cupboards underneath. Vices and other tools may be mounted on the benches.

Where the work is solely unit based – such as overhauling gearboxes – the benches may be aligned in rows separate from the vehicles – with the use of

Table 4.2 Tyre Markings and Wheel Serviceability

No	Task	Detail	Typical answers	Action points
1	Use necessary PPE	Mechanic's gloves	Protects hands from cuts and burns when handling tyres	Look out for hot tyres and sharp edges
2	Correctly identify relevant tyre data for vehicle	Size, pressure, type, fitment	195/55 R 19 2 bar (30 psi) M&S Symmetrical	Use either vehicle mfg or tyre mfg data sheets
3	Locate and identify tyre markings	Tyre and wheel size, speed rating, load index, tread wear indicator, aspect ratio	195/55 R 19 82 H	Sketch tread wear indicator between tread ribs
4	Measure tread depth	Use MOT green depth gauge	Legal minimum 1.6 mm	Check across full width of tread and all circumference
5	Examine tyre condition	Look for damage to the tyre	Cuts, lumps, bulges, tears, abrasions, intrusions, movement on rim, concussion, tread separation	Check the tyre pressure
6	Examine tread wear patterns	Look for uneven wear (see 7), skid wear and patch wear on tread	Skidding, out of balance	
7	Identify reasons for abnormal wear patterns	State causes of edge or wear at one point	Incorrect tyre pressures, steering misalignment	
8	Examine wheel condition	Look for damage to wheel	Impact damage, cracks, distortion, run-out, security	
9	Check valve condition and alignment	Is it in straight?	May be damaged or bent after impact	
10	Record faults			List wheel and tyre faults
11	Complete data collection sheet			Record findings on data collection sheet

Figure 4.9 Small socket set.

stands, or rigs, for the gearboxes or other major components. The drawers, or open racks, will then contain the special tools needed for the job in hand.

Fabrication area – this is where items are fabricated and welded. It usually contains rollers, bending machine, croppers and MIG or TIG welding equipment. Specialist trained engineers in this area will provide these services to enable you to carry out your overhaul and repair tasks.

Composites shop – where composite components are manufactured or repaired – a specialist clean area. This shop is staffed by specialist technicians who will make new parts or carry out specialist repairs to enable you to overhaul the motorsport vehicle.

Design studio – where vehicles and components are designed and modified. Usually using computer-aided engineering (CAE) – that is the computer-aided design (CAD) is connected to the computer-aided manufacture (CAM) machine tools such as the five-axis milling centre. You will find the design staff supportive in providing technical data for your overhaul procedures.

Model shop – named because they create models, or macquettes, of vehicles for design and testing purposes – including scale models for the wind tunnel – as well as specialist full-scale parts such as aerofoil wings. The model shop is both a source of data and specialist parts and skills. The model shop in larger, or older, firms may incorporate clay and/or wood handling equipment and skills.

Table 4.3 Pre-event Set Up – In Workshop

No	Task	Detail	Typical answers	Action points
1	Use necessary PPE	Mechanic's gloves, goggles		
2	Obtain vehicle set-up data	Check with appropriate authority	Use set-up management system	This may need a signature
3	Check previous data and driver comments		Use set-up data log	
4	Check battery serviceability	Use battery test procedure		System will not operate without a sound power supply
5	Check oil and fluid levels	Note varianc		
6	Check and adjust suspension and corner weights	Refer to data	Complete set-up documents showing changes	
7	Check and adjust steering geometry	Castor, camber, SAI, toe-out on turns	Complete set-up documents showing changes	
8	Check brake conditions	Measure pads, discs	Record on inspection sheet	
9	Check driver safety equipment	Seat belt harness, seats, roll cage, padding and mountings, fire extinguisher system		
10	Carry out spanner check	Check against checklist. Rectify as needed	Note any loose, damaged, or broken fastenings	
11	Record faults		Use set-up data file	Record any system faults
12	Complete data collection sheet			Record findings on data collection sheet

Paint shop – where the vehicles are painted. Again this shop has specialist staff and equipment. The overhauled vehicle, or its panels, will be refinished in this shop. An interesting move in refinishing is to use vinyl film instead of paint; the US army dragster, which is probably the fastest race car ever

Figure 4.10 Torx drive set.

Figure 4.11 Allen key with handle.

Table 4.4 Pre-event Inspection – At Event

No	Task	Detail	Typical answers	Action points
1	Use necessary PPE	Mechanic's gloves, goggles		
2	Check previous data and driver comments		Use set-up data log	Check previous data and driver comments
3	Check tyres	Fitment (direction/ corner), pressure, type, compound	Use set-up data log	
4	Check oil and fluid levels	Note variance	Check oil and fluid levels	Note variance
5	Check and adjust steering geometry	Castor, camber, SAI, toe-out on turns	Complete set-up documents showing changes	
6	Check security of bodywork	Panel attached, clips closed, screen clear, decal correctly located	Use checklist	May choose to photograph for record
7	Check seat belt harness with driver in place	May need adjusting		Note any pre-sets
8	Arm fire extinguisher system	Check setting, switches		
9	Record faults		Use data log	Record any system faults
10	Complete data collection sheet			Record findings on data collection sheet

made, uses vinyl film. They applied this in their workshop at the Indianapolis raceway, known as the Brickyard. The comment of the technician applying the material – which is printed and cut on site – was, 'It's lighter than paint.' When it is on, it is almost impossible to tell that it is not paint.

Parts and storage – the safe and secure storage of parts, both new ones and ones waiting for completion of the overhaul, is very important. Race car parts cost from 10 to 100 times the equivalent of the road-going equivalent.

Dynamometer and test shop – referred to as the dyno, or the rolling road. This allows the road wheels to sit on the dyno rollers and the power and torque measurements to be taken. Running any competition engine makes a lot of noise – so this is usually situated separate from the main building, or suitably noise-insulated from it.

Figure 4.12a Locking wire.

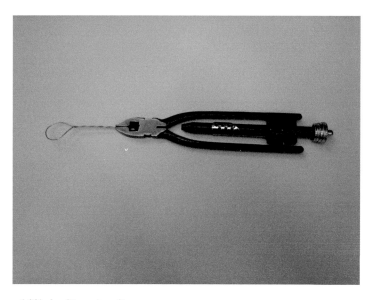

Figure 4.12b Locking wire pliers.

Figure 4.13 Portable vehicle repair unit.

If it is just the engine that is being tested, then a test cell is used. A test cell, of course, requires much less space than a rolling road. This takes the form of the engine mounted on a frame that is attached to the dynamometer (dyno). Because of their exceptional noise, aircraft engines are tested underground; some race teams do this too.

Tools and Equipment

Table 4.5 lists the equipment needed for the machine shop. In this section we will look at the fuller range of equipment that is used in vehicle and unit overhaul.

Data

As a technician in motorsport you will have to collect, collate, store, use and communicate data.

Contrasting with the Race Car Engineer is the Cycle Mechanic

Dr Bike – This is a concept used by a number of cycling organisations in an attempt to make cycling safer. Sometimes it is combined by police post-code

Table 4.5 Workshop Equipment

No	Item	Specification	Purpose	Comment
1	Small lathe	150–200 mm swing with 450–600 mm between centres	Making small items such as spacers, and cleaning up round parts	This will need a range of tools and chucks
2	Off-hand grinder – small	Approximately 150 mm diameter wheels – 1 fine, 1 coarse	General sharpening and cleaning	
3	Off-hand grinder – large	Approximately 250 mm diameter wheels	Sharpening drills and tools	Keep flat for accurate work
4	Pillar drill	5 mm chuck and variable speeds	Variety of drilling	Need variety of vices, or clamps, and drills and countersinks
5	Band saw	Approximately 100–150 mm cut	Cutting up steel stock	
6	Hydraulic press	20 tonne	Removing and replacing bearings and pins	
7	Milling machine	5-axis CNC milling centre	Manufacturing small parts	Used in conjunction with CAD system
8	Buffing – polishing	Floor-mounted buffing and polishing heads	Finishing parts	
9	Parts cleaning bath	Chemical cleaning bath with pressure spray	Cleaning parts	

marking bicycles to reduce crime. A number of bicycle repair companies also run Dr Bike sessions, as well as raising safety awareness it can generate extra business. If a session is run correctly it can be a good public relations exercise, give safety awareness, spread bicycle knowledge and help to generate business. Visits to school and shopping malls appear to be the most successful way of running these events.

Checklist – this checklist is broken down to cover the main areas of a Dr Bike session, answers are yes or no. If no, the point would be further investigated and the owner advised appropriately. A Dr. Bike check should take about half an hour.

Preventative Maintenance

Like a car, servicing the bicycle will prevent breakdowns and accidents. The usual method is like car servicing too, what is called menu servicing, a number of packages offered for fixed prices. These usually take the form of

Table 4.6 Overhaul Equipment

No	Item	Purpose	Note
1	Four-post hoist – with wheel-free adaptor	Raise vehicle and independently raise one corner	MOT compliant
2	Two-post wheel-free hoist	Lift vehicle under chassis so all wheels are free	
3	Body scanner	Scanning bodywork to produce CAD drawings	Good when rebuilding historic vehicles
4	Cam profiler	Regrinding camshafts	Make cam profiles to your design
5	Castor, camber and SAI (KPI) gauges	Checking steering geometry	
6	CMM machine	Accurately measuring components for CAD drawings, or reverse engineering	
7	Coil spring gauge	Testing coil spring rate	Check all four springs
8	Compression gauge	Testing engine compression pressure	Compare reading of each cylinder
9	Corner weights	Checking weight distribution at each corner of the vehicle	Adjust suspension and redistribute weight as needed
10	Crankshaft grinder	Regrinding crankshafts	
11	Dial test indicator (dial gauge)	Measuring movement – such as for valve lift	
12	Durometer	Measuring the hardness of tyre treads	Check temperature first
13	Engine boring equipment	Re-boring cylinder blocks	
14	Granite table	Providing a smooth and level surface on which to set up the vehicle	Cost is up to £1m, used for world class vehicles
15	Horizontal milling machine	Milling surfaces – such as cylinder head faces	
16	Laser, or light, suspension aligning gauges	Checking steering and suspension alignment	The manual system can achieve the same results
17	Mercer gauge	Measuring the diameter of a cylinder bore	
18	Micrometres – range, internal and external	Measuring inside or outside surfaces – such as cylinder bores and crankshaft bearings	

(continued)

Table 4.6 Cont.

No	Item	Purpose	Note
19	Pressure washer	Cleaning mud off the vehicle	Ensure EPA compliance when used
20	Surface grinder	Grinding surfaces such as cylinder head faces	
22	Turn tables	Measuring toe-out on turns and carrying out steering checks	Used in conjunction with No 5
23	Tyre machine	Removing and refitting tyres	Special machine needed with aluminium alloy rims
24	Tyre pressure gauge	Measuring tyre pressures	Need accurate gauge on motorsport vehicles – check temperature before adjusting
25	Tyre temperature gauge	Measuring tyre temperature	Take three measurements on each tyre – inside, middle, outside of tread
26	Vertical milling machine	Milling tasks such as when enlarging inlet ports	
27	Welding equipment (MIG or TIG)	All kinds of joints and repairs	
28	Wheel balancer	Static and dynamic wheel balancing off the vehicle	Use only approved weights – usually stick-on on inside of rim

Table 4.7 Pre-overhaul Data Check

No	Data	Source	Comment
1	Test data	Report on test	This should be added to vehicle log
2	Unit removal data	Workshop manual	
3	Unit test data	Unit manufacturer	
4	Unit stripping data	Unit manufacturer	
5	Equipment operation data	Equipment manufacturer	
6	Replacement parts	Parts manual	Parts used should be recorded
7	Unit assembly and test data	Unit manufacturer	This should be added to vehicle log
8	Set-up data	Vehicle set-up log and/or unit manufacturer	This should be added to vehicle log

Table 4.8 Dr Bike Check List

No.	Item	Specific	Yes/No
1	Front wheel	True (not buckled). No broken/missing spokes. Good rim	
2	Front tyre	Good tread. No splits. cracks or holes. Pumped hard. Valve straight	
3	Front hub	No wobble. Turns smoothly. Wheel securely fixed	
4	Front mudguard	Firmly fixed. No sharp mudguard stays	
5	Front brake blocks	Correctly fitted. Not worn away	
6	Front brake	Firmly fixed. Correctly adjusted	
7	Front brake lever	Comfortable position. Firmly fixed. Cable not frayed	
8	Headset/steering	No wobble. Correctly adjusted	
9	Handlebars	Not distorted. Ends protected	
10	Front forks	Appear true and undamaged	
11	Frame	Appears true and undamaged	
12	Rear brake lever	Comfortable position. Firmly fixed. Cable not frayed	
13	Rear brake	Firmly fixed. Correctly adjusted	
14	Rear brake blocks	Correctly fitted and aligned. Not worn away	
15	Rear mudguard	Safely fixed. No sharp mudguard stays	
16	Rear tyre	Good tread. No splits. cracks or holes. Pumped hard. Valve straight	
17	Rear wheel	True. No broken/missing spokes. Good rim	
18	Rear hub	No wobble. Turns smoothly. Wheel securely fixed	
19	Bottom bracket	No wobble. Lock ring tight. Sufficiently lubricated	
20	Pedal cranks	Straight	
21	Pedals	Complete. Turning freely. Not bent	
22	Chain wheel	Not bent. Teeth not worn	
23	Chain guard	Firmly fixed. Not bent	
24	Chain	Not too worn. Not slack. Lightly oiled not rusty	
25	Gears	Properly adjusted. Lubricated sufficiently	
26	Saddle	Safely fixed. Straight. comfortable height (unless BMX)	
27	Rack/carrier/bags etc.	Firmly secured	
28	Front lamp (if carried)	White. Firmly fixed. Good light to front	
29	Rear lamp (if carried)	Red. Firmly fixed. Visible to rear	
30	Reflectors	Clean and secure	

Table 4.9 Service Menu

No	Tasks	Bronze	Silver	Gold
1	Safety check of complete bicycle	X	X	X
2	Brake balancing and servicing	X	X	X
3	Clean or replace chain	X	X	X
4	Fit new tyres/tubes if needed	X	X	X
5	Service or replace brake cables		X	X
6	Service or replace gear cables		X	X
7	Check and fine tune gear change		X	X
8	Service or replace cassette/sprocket		X	X
9	Clean and check/replace chainset		X	X
10	Clean and check/replace pedals		X	X
11	Inspect/replace bottom bracket			X
12	Inspect and adjust headset			X
13	Check and true wheels if needed			X
14	Service front hub			X
15	Service rear hub			X
16	Replace bar tape/grips			X
17	Waterless clean			X

basic, middle and best – in the example labelled bronze, silver and gold. The pricing is based on the time and effort involved for each, typically this will be bronze half hour, silver 1 hour and gold 2 hours. Parts will be charged at normal retail price, so adding to the profit margin for the job. For example, if a chain is replaced on the bronze service this will take less time than cleaning the original one and give a profit on the sale of the new chain. Both customer and repair shop win, everybody is happy.

Risk assessments – these are essential for all tasks, see Health and Safety chapter.

SOP – standard operating procedures are covered in the chapter of that title.

Skills and Questions

1. Probably the most important task is to try work experience in the area of MIR, preferably with more than one company, or in more than one department in a big company.
2. Make a list of tools and equipment to be found in any type of maintenance workshop which interests you.
3. Using a workshop manual for any type of vehicle that you are interested in, look up the service schedule – list the tasks required and note the service intervals.

4. Three dispensing/vending machines which are very common are: petrol pumps, snack vending machines and photocopy machines. Look at the regulations, installation requirements and maintenance needed for one of them. Tip – your school/college estates department will probably be delighted to help you.
5. Carry out a *Dr Bike* check on a bicycle or scooter, note your findings.

Chapter 5

Manufacturing, Processing and Control Practices

Industry across the world is classified into three major sectors, these are:

1. **Primary** – the economic activities usually depend on the environment of that specific region. The economic activities in a primary industry revolve around the usage of the natural resources of the planet like vegetation, water, minerals. In this industry, the major economic activities are harvesting and hunting, fishing, mining, agriculture, extraction and forestation. In relationship to engineering, the extraction of both iron ore and bauxite is very important to provide the required basic materials. Engineers are important in making, installing, servicing and repairing the machinery and tools to do this primary sector work.
2. **Secondary** – the economic activities revolve around adding value to the natural resources by transforming the various raw materials into usable and valuable products. This is done via several processes, manufacturing and construction industries. Again engineers are important in making, installing, servicing and repairing the machinery and tools to do this secondary sector manufacturing work.
3. **Tertiary** – the major economic activities include exchange and production. Production usually involves the provision of a large array of services consumed on a large scale by millions of consumers. When we talk of exchange, this involves transportation, trade and communication facilities that are often used to overcome distances. The tertiary sector is also referred to as the service sector. The service sector sells, maintains and repairs the manufactured goods. Engineers form a major part of this tertiary, or service sector, installing, maintaining and repairing machines, mechanical and electronic equipment, cars, trucks, boats and aeroplanes.

DOI: 10.1201/9781003284833-5

Tech note

Let's take an example, the primary sector mines the materials from the ground and converts them into iron and aluminium ingots – pieces of useable metal for car makers. The secondary sector makes them into useable items, maybe engines for cars. The tertiary sector sells, maintains and repairs the cars made by the secondary sector from primary materials.

Historically – in the early days of civilisation a person would dig up the copper and tin, melt them together to form bronze, then make items from the bronze and sell them. This means that that person was working in, in effect, all three sectors. Of course, production was very low and very slow. As civilisation progressed different groups of people were involved in the different processes. Some mining the metal, some smelting the bronze, others crafting the objects.

The invention of the steam engine and electricity meant that production could move into large-scale factories. So manufacturing could be further divided into smaller jobs. The machines doing a lot of the work much faster and more accurately than any person.

At the same time small specialist businesses were being developed, such as wheel makers, bobbin makers, blacksmiths and nail makers. These manufacturers bought in part processed materials and using hand skills converted them into useful products.

The advent of computers since about 1970 has led to computer-aided engineering – **CAE**. This has allowed complete automation of many production processes. The interconnectivity of design computers – **CAD** – and computer-aided manufacturing – **CAM** – has led to the term **CADCAM**.

However, not all components are readily made by machinery, and the costs of machinery often outweigh that of the finished products, so most engineering companies operate on a mixture of both machine and person production. This often includes the person carrying out quality checks too.

Terms Used to Identify Production Methods

Mass production – this term is used to describe items made in very large quantities. Items made in millions. The mass production of products may involve buying in some parts, or partially completed items, from other suppliers and processing them, or assembling them to make a new component. Examples of mass-produced items are: kettles, pans, washing machines and cutlery – in other words most consumer goods. Mass production may involve other types of production, it is an overriding term for the production of goods in large numbers, it is not a single production method.

Flow production – also referred to as **flow-line production**. This is the production method developed by Henry Ford for the production of cars and other vehicles. The vehicle, or other item, progresses through what are called **stations**, places where the items stops to be assembled, then moves to the next station. For example, in car manufacturing, at one station the engine is fitted to the car shell, at the next station the suspension is fitted. JIT is used to ensure that the correct engine for the particular car is available so that a constant flow speed is maintained. Typical car production from the arrival of the basic body, fitting all the body part, painting, then fitting all the other parts, finally filling it with fuel and actually driving it off the production line takes about 24 hours. Mass-produced cars flow off such a production line at the rate of one per minute, these production lines are working 24 hours each day. For high-quality cars, the production techniques are very similar, although they may have less assembly stations and only produce five or six cars each day – usually only working eight-hour days.

Continuous flow production – this usually refers to production where time and temperature or heat, are important parts of the process. It involves exact timing for each part of the process. The production of sheet metal is an example, the molten steel is poured to form ingots, before they cool they go to the rolling mill to become sheet metal. The rolling process can only be carried out at a certain temperature, so the time between ingots and rolling is very limited. Another example is pizza production – this is very time and temperature dependent. The mixing of the dough, the cooking of the vegetables and meat, the initial cooking, the freezing and the packaging. In London, 250,000 pizzas are eaten each day, in America they eat 3,000,000,000 – 3 billion – each year.

Intermittent flow production – items from the rubber industry and the petrol chemical industry are usually in continuous production like steel products, they do not stop for holidays as it is expensive, if not impossible, to stop the flow. Other items may be made by intermittent flow. That is, production is run to produce enough goods to meet an order, or create a stock of items. This may be better illustrated by considering newspapers. The printing machines for newspapers run through the night to have the newspaper ready for the early morning. They then stop whilst the next days' news articles are written, to start the next evening.

Batch production – this is where a batch, or specific number, of articles is produced. The batch number will vary, and the identification of the batch may vary too. For special events, such as a major football match, Olympics Games or a Royal Jubilee, companies often produce special editions or limited editions of otherwise standard articles. Items as varied as cameras and cars, bearing a special logo, or colour, or specific shape. These are examples of batches. Batch production is also sometimes used to test the market. Other times it is that customers order in batches to satisfy their production schedules.

Jobbing shops – these tend to be specialist, high-quality, highly skilled, manually run workshops. They take on one-off and low-number production jobs. This title covers: precision engineering, lapping and grinding engineering, welding specialists, forge and foundry work, and many others too. Specialist vehicle restorers and race car engineers use jobbing shops and sometime do jobbing work – such as making a car body part for another restoration company. This is also a big area of employment for structural steel workers used in construction engineering. A lot of structural steel is cut and assembled on site – in effect jobbing.

Push production and **pull production** – these are two term which are often used in factories to describe the current situation, or focus. They indirectly indicate the stress level of the workers, and the machines. Push refers to making items that a company thinks will be sold, it is an abbreviation for pushing into the market. Walmart, the gigantic supermarket chain pushes items in to its stores based on what it thinks will sell. Pull refers to making items to meet a customer demand – this demand is usually critically time limited. The pull situation in manufacturing is much more demanding, hence higher stress levels.

Manufacturing cells – this refers to a system of manufacturing, usually of a high-precision nature, where specific processes are carried out in an enclosed area – a cell. The cell may be temperature controlled and equipped with extractors to remove all dust and contamination. Temperature changes cause metal to expand and contract, by maintaining a constant temperature any measurements are free from temperature change variations. Being dust-free prevents any contamination and damage to components. In car manufacturing, cells are used for tasks such as engine assembly, where tolerances are very fine. Once a cell is designed for a job, and proved in production, it can be replicated elsewhere. Car manufacturers have identical cells in different production plants. This also has the advantage that an engineer doing a task in one country, say England, can be sent to do the same task in another country, for instance Germany, without any need to understand the other language. Manufacturing cells are used in conjunction with JIT and other Japanese-based manufacturing techniques.

Production control – the job of a **production controller** is a senior post in many engineering companies, this job entails:

- Ensuring that all tools and equipment are ready for use, safe to use and are accurate.
- Preparing schedules for the jobs and processes, giving accurate timing, down to the second if needed.
- Ensuring the staff are properly trained and understand the full process.
- Works with quality inspectors and the quality circle team.

- Ensuring that the materials and components needed for production are available as needed.
- Being responsible for the use of energy and other resources.
- Being responsible for waste and re-cycling.

Extended product responsibility – EPR, to be borne in mind for all production are the EPR regulations. Before making anything, the questions of "what will be its life cycle?", "how will it be re-cycled?" "or scrapped?" must be addressed.

Skills and Questions

1. Research your local area and list the engineering companies, identify the type of work that they do.
2. Looking around your home, list the engineered items and briefly describe how you think that they were made.
3. State why the job of a production controller is important.
4. What is meant by EPR?
5. Describe the difference between push and pull production.

Chapter 6

Engineering Calculations

Introduction

In engineering we make a lot of calculations based on mathematical principles, that is we apply numbers to situations, this is to guide us engineers on how to tackle a job, or work out how to go about making something. It is about calculating sizes, shapes, weights, loads, velocity and acceleration. This chapter sets out to take you through typical formulas and concepts of engineering calculations.

Do not worry about how the calculations work, when you work through the course you will apply numbers to the formulas and get answers, when you have completed the course you will have used most of the methods and formulas used in everyday engineering.

Tech note

To gain the award of a full T Level qualification it is necessary to achieve a separate Level 2 qualification in mathematics by the time that the T Level is completed. This means either GCSE Grade 4 or higher in mathematics; or a Level 2 Functional Skills qualification, or equivalent, in mathematics.

Calculator – as an engineer a good calculator is an essential work tool for doing the job. It is also essential for your tests and examinations. Using a calculator shows a professional approach to your work, NOT a mobile phone. The Casio range of calculators all use the same logical entry system, similar to the way that you would write calculations on papers.

Estimations or rough checks – when carrying out any calculation it is good practice to do a mental, or rough pencil calculation, as to the order of the likely outcome.

DOI: 10.1201/9781003284833-6

Tech note

One estimation which always makes the author smile is in the film *The Italian Job*, the scene where they are loading gold bullion into the boot of a Mini. The density of gold is 19,300 kg/m³ so one ingot of gold is difficult to lift, a layer of them on the boot floor would crush the car.

Knowing your multiplication tables is helpful for estimating, as is using your engineer's note book.

BODMAS and Transposition of Formula

It is expected that you will have undertaken a course of study for GCSE or Functional Skills in Mathematics before starting this T Level; but we'll begin with a re-cap of work to date. The two main keys to carrying out engineering calculations are BODMAS and Transposition of Formula.

BODMAS – is about the order in which you carry out your calculations. Think of it in the same way as making a cup of tea. Get cup, place tea-bag in cup, add boiling water, stir, and add milk. Tea will taste great; change the order of any of the actions and it will awful, or even just a mess on the table top.

> Always tackle calculations from left to right in the order of **BODMAS** if one of the terms is not present, move on to the next term in order.
>
> **B** is for brackets – do the calculations inside the brackets so that they are not needed anymore.
> **O** is for order of powers or roots – raise the number to the indicated powers, or find the roots, as needed.
> **D** is for division – divide where needed.
> **M** is for multiplication – multiply as needed.
> **A** is for addition – add as needed.
> **S** is for subtraction – take away as needed

As you get proficient, you'll see that you can frequently cancel out numbers in formulas, especially if you learn your multiplication tables and revise your calculations with fractions. You should look out for number patterns and recurring themes in your calculations.

Transposition of formula – is about balancing both sides of the equals sign. Think of the equals sign (=) as the pointer on old-fashioned scales, indication balance. That is, both sides of the equation are of equal value. So, if you change a value on the left, you must compensate it on the right by the same amount. The trick is that if you remove a positive value figure from the left, you need to add the same figure as a negative value to the right. In the same way that a divisor on the left becomes a multiplier on the right. Of course, this applies in reverse.

No matter how large, or how small, the calculation is, the rules for BODMAS and the transposition of formula always apply.

Tech note

It is a good idea to keep a **pocket-size notebook** to record important calculations and other information, this allows you to look back and compare information such as dimensions and cutting speeds. Some space engineers probably wish that they had re-read their notes when the $500,000,000 space rocket Ariane 5 blew up 40 seconds after launching at an altitude of 3,700m (12,300 ft). The problem was a decimal point error in the software calculations.

Indices or Powers

If we write a number such as 5^5 we say it is five to the power of five. The big five is called the base number, the little five (^{5}) is called the index, and the plural of index is indices. In normal speaking, indices is often used for both singular and plural.

$$5^5 = 5 \times 5 \times 5 \times 5 \times 5 = 3125$$

For another example:

$$3^2 = 3 \times 3 = 9$$

In algebra when we are working out a calculation, we often use a letter. For example:

For a^3 we say a to the power of three. In other words:

$$a^3 = a \times a \times a$$

In some calculations we do not know the value of the index, it is given a letter, for example:

$$5^d \text{ or maybe } a^x$$

Multiplication

When you multiply numbers of the same base with different indices, for example:

$$3^2 \times 3^5 = (3 \times 3) \times (3 \times 3 \times 3 \times 3 \times 3) = 3^7$$

In other words, add the indices

$$3^2 \, 3^5 = 3^{2+5} = 3^7$$

When two or more numbers with the same base are multiplied, add the indices.
The general case is expressed as:

$$a^m \times a^n = a^{m+n}$$

Tech note

In mathematics some terms are said in a different way to others of the same type. For instance 3^2 is said as 'three squared' not three to the power two. And 4^3 is said as 'four cubed' not 'four to the power of three.'

Division

This is the same rule but it uses a negative instead of the positive.

$$\frac{a \times a \times a \times a}{a \times a} = a^{4-2} = a^2$$

So

$$\frac{1}{a^2} = a^{-2}$$

And

$$\frac{a^x}{a^y} = a^{x-y}$$

$$\frac{a^x}{a^x} = 1 = a^{x-x} = a^0$$

$$a^0 = 1$$

And

$$\frac{1}{a^x} = a^{-x}$$

Powers

The same rules

$$a^x \times a^y \times a^z = a^{x+y+z}$$

Roots

Roots is another variation on indices.
The square root of 4 is written $\sqrt{4}$, this is 2.

$$2 \times 2 = 4$$

$$4^{\frac{1}{3}} = \sqrt[3]{4} = 1.587$$

The rule is

$$a^{m/n} = \sqrt[n]{am}$$ should read n root of a to the power of m.

Circles

Circles are frequently measured in degrees:

One revolution of a circle = 360°

A circle may be divided into smaller parts, these smaller parts have the same name as the fractions that we use on a clock face, these are **minutes** and **seconds**

1° = 60 minutes
1 minute = 60 seconds

The usual abbreviations are min and s; sometimes one or two small dashes – like feet and inches symbols – are used. That is $1° = 60' = 3600''$.

Laws of Indices

Rule	Example
$a^m \times a^n = a^{m+n}$	$2^5 \times 2^3 = 2^8$
$a^m \div a^n = a^{m-n}$	$5^7 \div 5^3 = 5^4$
$(a^m)^n = a^{m \times n}$	$(10^3)^7 = 10^{21}$
$a^1 = a$	$17^1 = 17$
$a^0 = 1$	$34^0 = 1$
$\left(\dfrac{a}{b}\right)^m = \dfrac{a^m}{b^m}$	$\left(\dfrac{5}{6}\right)^2 = \dfrac{25}{36}$
$a^{-m} = \dfrac{1}{a^m}$	$9^{-2} = \dfrac{1}{81}$
$a^{\frac{x}{y}} = \sqrt[y]{a^x}$	$49^{\frac{1}{2}} = \sqrt[2]{49} = 7$

Figure 6.1 Laws of indices.

In more advanced calculations it is usual to use the measure of angles in radians. This is used as it makes more complex calculations easier. It is easier because it relates the angle, or part of the circumference of a circle to the radius.

Circumference of circle = πD

that is $\pi \left(\dfrac{22}{7}\right)$ x diameter

Or $2\pi r$

that is 2 x π x radius

A radian is the angle subtended by one radius on the circumference.

2π radians = 360°

$1 \text{ rad} = \dfrac{360}{2\pi} = 57.3°$

Tech note

Be careful, the 57.3° figure, like the 3.142 for π, is not an exact figure and so is not suitable for precision work.

$$1° = \frac{2\pi}{360} = 0.0175 \text{ rad}$$

A radian can be used to divide a circle into sectors, if we call the length of the radian on the circumference s, then

1 radian s = r
2 radian s = 2r
3 radian s = 3r
Φ radian s = Φr
The area of a circle = π r²
The area of the sector is calculated by

$$\text{Area} = \pi r^2 \text{x} \frac{\Phi}{2\pi}$$

$$= \frac{1}{2} r^2 \Phi$$

Triangles

Pythagoras theorem – the square on the hypotenuse is equal to the sum of the squares of the other two sides.

If we call the hypotenuse A and the other two sides B and C. The hypotenuse is the long side which is opposite the right angle in the triangle.

$$A^2 = B^2 + C^2$$

Therefore

$$A = \sqrt{A^2 + B^2}$$

This is sometimes referred to as the 3, 4, 5 rule as

$$5^2 = 3^2 + 4^2$$

$$25 = 9 + 16$$

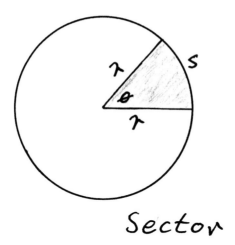

Sector

Figure 6.2 Sector.

Tech note

The 3, 4, 5 rule is very useful if you need to make something and you have not got access to much equipment, such as if you are building emergency accommodation in a disaster area with one of the engineers without borders organisations, all you need is a tape measure.

Trigonometrical (Trig) Functions

The inside angles of any triangle add up to 180°. With a right-angle triangle one angle is 90° and the sum of the other two adds up to 90°. So if you change the length of any one side, or the angle of any of the two non-right angles, the length of the other sides, or the other angle will change. These changes follow the rules of the trig functions. These are:

Sine, Cosine and Tangent

An easy way to remember the formulas for them is using the mnemonic **SOHCAHTOA** (see Figure 6.3).

As you might expect, as these ratios change with a specific function, they are all related, for any angle of A

$$\text{Tan } A = \frac{\text{Sin } A}{\text{Cos } A}$$

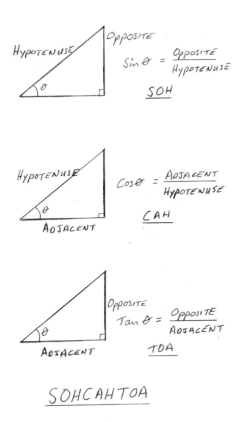

$$\sin\theta = \frac{Opposite}{Hypotenuse}$$

$$\underline{SOH}$$

$$\cos\theta = \frac{Adjacent}{Hypotenuse}$$

$$\underline{CAH}$$

$$\tan\theta = \frac{Opposite}{Adjacent}$$

$$\underline{TOA}$$

$$\underline{SOHCAHTOA}$$

Figure 6.3 SOHCAHTOA.

And for any triangle

$$\frac{a}{\sin A} = \frac{b}{\sin B} = \frac{c}{\sin C}$$

Graphs – these are very important in engineering as they are used for visual comparisons. Frequently, different test results are shown on the same graph for accurate comparisons. The general formula for a graph is:

$$Y = mX + c$$

Y is the value on the vertical axis of the graph, also called the dependent variable, X is the independent variable on the horizontal axis, m is the gradient of the graph, and y is where the curve intersects the Y axis. The gradient of the curve is in fact a tangent – opposite over adjacent – you will find this useful to know in harder calculations.

Tech note

With reference to graphs, the plotted line is called a curve, even if it is a straight line. The curve should be drawn as a best fit connecting the plotted points, a smooth curve or straight line, not joining the plotted points.

Simultaneous Linear Equations

Particularly in electrical and electronics work you can take a number of readings and finish up with two unknown equations. What you are trying to do is express the values of x and y in terms of each other. You are looking for an answer of the type x = 3 when y = 4.

There are three main ways of solving simultaneous equations, these are:

- by substitution
- by elimination
- by use of a graph.

Substitution – that is substituting one of the values, that is x or y, with the equivalent in the other term. In the following we first substitute the y value in equation (2) with the x value from equation (1).
Example

$$2x + y = 10 \tag{1}$$

$$3x + 2y = 17 \tag{2}$$

Taking equation (1) in terms of y

$$y = 10 - 2x \tag{3}$$

Replace y in equation (2) with its value of $10 - 2x$

$$3x + 2(10 - 2x) = 17$$

Therefore x = 3
Replace x in equation (1) with 3

$$(2 \times 3) + y = 10$$

Therefore y = 4

<u>Answer x = 3 when y = 4</u>

Elimination – that is adding or subtracting one of the equations from the other, to eliminate either the x or the y term. To do this we may need to multiply or divide one of the equations.

Example

$$3x + 4y = 11 \tag{4}$$

$$x + 7y = 15 \tag{5}$$

Multiply equation (5) by 3

$$3x + 21y = 45 \tag{6}$$

Subtract equation (4) from (6)

$$17y = 34$$

$$y = 2$$

Substitute y = 2 in equation (4)

$$3x + (4 \times 2) = 11$$

$$3x = 11 - 8$$

$$x = 1$$

<u>Answer x = 1 when y = 2</u>

Use of Graphs

The use of graphs is quite simple, all that you do is plot three points for each of the equations, where the points intersect is the solution of the type x = ? when y = ? (see the graph in Figure 6.4).

Quadratic Equations

Quadratic equations, as the name suggests, are raised to the power of two, that is instead of the unknown being x as in a simultaneous equation, the unknown in a quadratic equation is x^2.

Quadratic equations are often used in calculations relating to velocity and light, which use square values.

The normal format for a quadratic equation is of the type:

$$3x^2 + 5x - 3 = 0$$

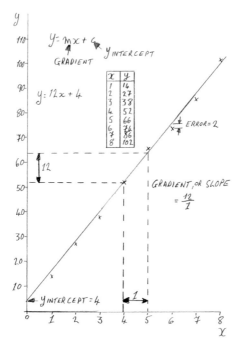

Figure 6.4 Use of graphs.

The standard form is written:

$$ax^2 + bx + c = 0$$

There are two ways of solving quadratic equations, these are:

- Factorisation
- Using a formula

Factorising a quadratic equation is the process of breaking the equation down into the product of its factors, this is the reverse of multiplying out. In all cases the equation must be organised so that one side is 0. For example:

$$x^2 + 3x - 4 = 0$$

is factorised to

$$(x - 1)(x + 4) = 0$$

Each of the factors are possible solutions:

x − 1 = 0 or x + 4 = 0

So either

x = 1 or x = −4

Sometimes, after a little experience, you can easily see the factors, or you can experiment. Sometimes you cannot easily find the factors, the solution then is to use the formula – see Figure 6.5.

Note that in the equation $3x^2 - 11x - 4 = 0$, the values for the formula are a = 3, b = −11 and c = −4

Data

Just about every individual and organisation collects data of some sort or another. At a simple level it may be checking the instruction sheet to see

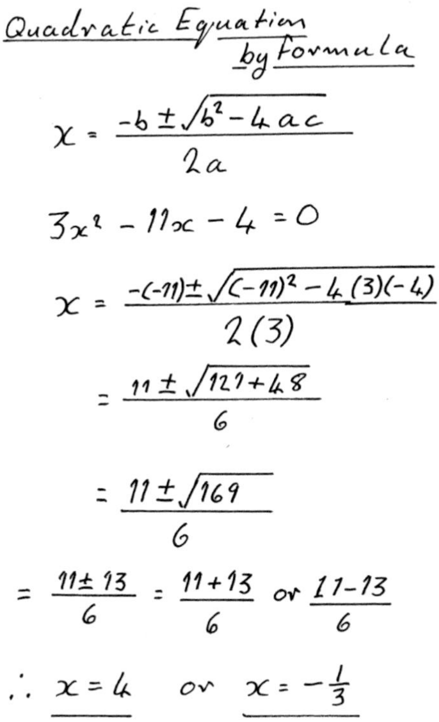

$$\underline{Quadratic\ Equation\ \underline{by\ Formula}}$$

$$x = \frac{-b \pm \sqrt{b^2 - 4ac}}{2a}$$

$$3x^2 - 11x - 4 = 0$$

$$x = \frac{-(-11) \pm \sqrt{(-11)^2 - 4(3)(-4)}}{2(3)}$$

$$= \frac{11 \pm \sqrt{121 + 48}}{6}$$

$$= \frac{11 \pm \sqrt{169}}{6}$$

$$= \frac{11 \pm 13}{6} = \frac{11 + 13}{6} \ or \ \frac{11 - 13}{6}$$

$$\therefore \ \underline{x = 4} \quad or \quad \underline{x = -\frac{1}{3}}$$

Figure 6.5 Quadratic by formula.

what the tyre pressures should be; or looking at the train timetable to find the next train to get to the office.

Tech note

Data, noun plural, facts and statistics collected together for reference or analysis. The singular is datum. Data includes statistics and information.

Engineers collect data on such things as strength of materials, size of components being machine made, cutting speeds, reliability of equipment, power output of motors, voltages and current used.

The Norm

The norm is something which is usual, typical or standard.

The secret in the use of data is actually knowing what they show and what you can do with that information. In other words, what is the norm and what are the extremes, what makes them these cases and how can we utilise this information. This concept of the norm came about through population studies – scientists visiting other countries many years ago and comparing the height and other attributes of the inhabitants of these countries. They would then say the norm for the height of people from country A is x so that they could compare them in country B which is y. So, they developed the concept of the normal distribution curve. That is to say, not all people in country A are the same height; but the bulk of the population will vary within a few centimetres either side of x. In addition, there will be both very tall people and very short ones – those outside the norm. If you go into any high street clothes shop, you'll find that they usually only stock a limited range of sizes – the ones which fit those within the norm. Often this range is simply small, medium and large in their own ratings and varies with manufacturers.

Understanding and Using Data

To use data, it is a good idea to consider **Bloom's taxonomy**. It states the six levels of the cognitive domain. Most people work in the lower three levels, these are **knowledge, comprehension** and **application** in Bloom's terminology. In other words, understanding the description of the data, being able to describe and discuss the actual content of the data and being able to apply the actual data in a real situation.

The three higher levels are **analysis, synthesis and evaluation.** Analysis means breaking the data down, understanding both the content and the structure. Synthesis is picking out the bits which you think were most important and re-mixing them to form a new structure. Evaluation is making a judgement on whether the new structure worked – then making more changes to improve it again. It is a continuous process.

Recording Data

Recording data is very important, how you record them will control the way in which you can use them. Businesses such as takeaway shops use data to control their opening hours to match their customer needs. For instance, rarely will you find a chip-shop open before 12 pm; but London kebab shops often stay open till 4 am.

Don't waste time on recording data which you will not use, and remember that if it is not recent it won't be relevant. Mood, fashions and tastes change quickly; as does technology and many other factors.

Tally Chart

A simple way of collecting data is the use of a tally chart, the first column has the options listed. The second colour is the tally, this is made in pencil line strikes up to four, then the fifth is across. The third column is the tally totalled.

Tally Chart for Colours at a show

Colour	Tally	Frequency
Red	⊥⊦⊦⊢	5
Orange	I I I	3
Yellow	I I	2
Green	I I I I	4
Blue	I I I I	4
Inigo	I I	2
Violet	I	1

If it is the number of attendees at an event a simple manual tally counter is ideal, push the lever for each head counted. Total number will be displayed.

Spreadsheets

Spreadsheets are a great way of recording data, and of course you can carry out detailed analysis, synthesis and evaluation using *what if* scenarios on

Figure 6.6 Manual counter.

them. That is, inserting projected values in the columns and seeing what happens to the results. This is of particular value when you are concerned with data changes of very small margins, say less than one percent. What the extra profit will be increasing the price of something by a few pence will make to the bottom line.

Presenting Data

Spreadsheets are good for the recording and analysis of data; but to get your message over, to both yourself and to others, a visual form is useful. Print them out and pin them up on the wall to help motivate yourself and your staff or colleagues.

Bar Charts and Stacking Bar Charts

The bar chart and stacking bar chart are good to show growth and changes in emphasis. The best approach is probably to make them in colour.

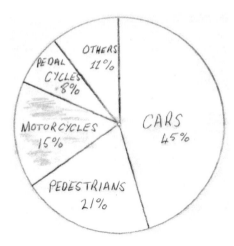

Figure 6.7 Pie chart.

Pie Charts

The pie chart is a really good way of showing percentage data as it gives a clear indication of percentages. A circle is made up of 360° so 10% is 36°. To find, for example, 30% as an angle for the pie chart:

30/100 × 360° = 108°

Normal Distribution

Calculations for normal distribution are simple to do, the following gives you two examples, ungrouped and grouped data use different methods. It is useful to be able to calculate the mean (average), the variance from the mean and the standard deviation.

Mean is another word for average, with the specific meaning of being calculated from the data by adding all the points together and dividing by the number of points.

Variance is another word for spread, the sum of the differences squared and divided by the number in the sample. The reason for squaring them is to remove any negative signs – a negative squared becomes a positive. **Standard deviation** is the square root of the variance.

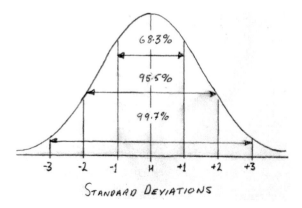

Figure 6.8 Normal distribution bell curve.

> **Tech note**
>
> 99.7% of normal data usually fall within six standard deviations. The quality control methodology Six Sigma is based on this concept of all manufactured items to fit within six standard deviations – the abbreviation for which is the Greek letter sigma σ.

Ungrouped Data

The following is a table of voltages measured across the lighting circuit of a low-voltage system in a workshop to check volt-drop.

Sample	Voltages	Differences from mean	Differences squared
1	119	1.46	2.13
2	120	0.46	0.211
3	120	0.46	0.211
4	120	0.46	0.211
5	121	0.54	0.291
6	119	1.46	2.13
7	122	1.54	2.37
8	122	1.54	2.37
9	123	2.54	6.45
10	123	2.54	6.45
11	119	1.46	2.13
12	118	2.46	6.05
13	120	0.46	0.211
Total: 13	Total: 1566		Total: 31.215

Mean:

$$\bar{x} = \frac{Total\ of\ voltages}{Number\ in\ sample} = \frac{1566}{13} = 120.46$$

Variance:

$$S = \frac{Sum\ of\ differences\ squared}{Number\ in\ sample} = \frac{31.215}{13} = 2.401$$

Standard deviation σ:

$$\sqrt{S} = sigma\ \sigma = \sqrt{2.401} = 1.549$$

Grouped Data

You are carrying out a quality check on suppliers. The table gives the sizes in mm of samples of 100 mm machine spindles.

Line	Range/ group mm	Mid- point x	Frequency f	fx	$\bar{x}$	$(x-\bar{x})$	$f(x-\bar{x})^2$
1	89–91	90	17	1530	99.07	−9.07	1398.5
2	91–93	92	18	1656	99.07	−7.07	899.7
3	93–95	94	19	1786	99.07	−5.07	488.4
4	95–97	96	20	1920	99.07	−3.07	188.5
5	97–99	98	30	2940	99.07	−0.07	34.3
6	99–101	100	50	5000	99.07	0.93	4326
7	101–103	102	35	3570	99.07	2.93	300
8	103–105	104	22	2288	99.07	4.93	535
9	105–107	106	18	1908	99.07	6.93	864
10	107–109	108	10	1080	99.07	8.93	89
			Total: 239	Total: 23678			Total: 9123.4

Mean:

$$\bar{x} = \frac{\Sigma(fx)}{\Sigma f} = \frac{23678}{239} = 99.07$$

Variance:

$$S = \frac{f(x-\bar{x})2}{\Sigma f} = \frac{9123.4}{239} = 38.17$$

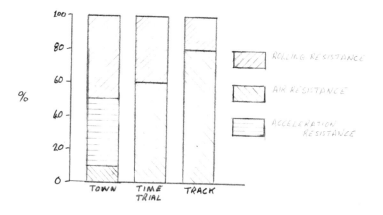

Figure 6.9 Stacking bar chart.

Standard deviation σ

Sigma $\sigma = \sqrt{S} = \sqrt{38.17} = 6.178$

Surface Area and Volume

The surface area of rectangular-shaped objects is simply:

Area = length × breadth

Area = l × b

for each face added together.
And the volume is calculated by:

Volume = length × breadth × height

Volume = l × b × h

For **cylinders, spheres** and **cones** the cal culation is slightly more complex.

Solid shape	Surface area	Volume
Cylinder	$2\pi rh + 2\pi r^2$	$\pi r^2 h$
Sphere	$4\pi r^2$	$\frac{4}{3}\pi r^3 h$
Cone	$\pi rl + \pi r^2$	$\frac{1}{3}\pi r^2 h$

Figure 6.10 Cylinder, sphere and cone.

To find the slant height of the cone (l) use Pythagoras theorem

$$l^2 = h^2 + r^2$$

$$1 = \sqrt{h^2 + r^2}$$

Other Number Systems

The number system which we use in everyday life, and is used throughout the world, which uses the digits 1, 2, 3, 4, 5, 6, 7, 8, 9, 0 is known as the Arabic Numeral System, and the digits are based on Arabic numerals. The use of 10 digits led to our current decimal system, deci means ten. So we use decimals and **standard form** to base ten.

For quantities, such as volts, we may give a prefix for example millivolt, which means one thousandth of a volt, or kilovolt which means one thousand volts. Not all of the numbers which we work with have prefixes and using them can cause difficulties, so we use standard form, this is simply whatever the number is, get the decimal point to be just behind one or two digits by using standard form, that is ten to a power. For example:

We could say:

10 000 volts or 10 kilovolts or 10×10^3 volts

With low voltages we could say:

0.010 volts or 10 millivolts or 10×10^{-3} volts.

Tech note

It is good practice to say numbers out loud and translate them between the different forms.

The number base which we use for our ordinary numbers is called the **decimal system**, or **denary numbers**, which is it is to a base ten. We may consider it in columns, each column to the left being ten times that of the other.

1000000	100000	10000	1000	100	10	1
10^6	10^5	10^4	10^3	10^2	10^1	10^0

This is all very well with our calculations using a hand-held calculator; but when we want to carry out really big and complex calculations, as in computer applications – at the rate of several thousand per second, we may use another system.

Binary numbers – this system uses only two digits, these are 0 and 1. The numbers are powers of 2 that is base 2.

128	64	32	16	8	4	2	1
2^7	2^6	2^5	2^4	2^3	2^2	2^1	2^0

Hexadecimal numbers – form another number system, this counts from 0 to F.

Decimal numbers	4-Bit binary numbers	Hexadecimal numbers
0	0000	0
1	0001	1
2	0010	2
3	0011	3
4	0100	4
5	0101	5
6	0110	6
7	0111	7
8	1000	8
9	1001	9
10	1010	A
11	1011	B
12	1100	C
13	1101	D
14	1110	E
15	1111	F

Calculus

If you have played or watched cricket, or indeed almost any ball game, you may have noticed that the ball does not travel in a straight line, it travels in a curve. The curve is not part of a circle, it changes its shape as the ball's velocity changes. A good fielder at a cricket match will look at the ball as it leaves the cricket bat, and estimate where it is going to land. The fielder does this from skill and experience. The ball falls as its velocity is reduced. It makes a curve in the air which reflects its rate of change of velocity, or kinetic energy. Differentiation is part of the calculus which seeks to define such rates of change. That is, it is used to calculate the differences. The other part of calculus is integration, this is the opposite of differentiation.

Tech note

There are volumes of books written on calculus since it was first discovered in about 1615, it is a subject of study in its own right.

Differentiation – you learnt in graphs that the gradient of the curve is a tangent expressed by the formula $\dfrac{opposite}{adjacent}$, when this is a straight line the gradient remains constant. When the curve is an actual curve, the tangent changes as its position changes against the curve. Remember with differentiation we are calculating change, specifically rate of change. For instance, car acceleration and the rate at which an aircraft climbs in the sky.

Differentiation formula – the differentiation is calculated by dividing the change in the y axis by the change in the x axis. This is usually written:

$$\frac{dy}{dx}$$

Tech note

d is often referred to as delta – the Greek letter δ.

As this has been done many times, so it is standard procedure to use the established formulas, some examples are:

When	Derivative
$y = x$	$\frac{dy}{dx} = 1$
$y = 3x^2$	$\frac{dy}{dx} = 6x$
$y = 4x^3$	$\frac{dy}{dx} = 12x^2$
$y = 5x^4$	$\frac{dy}{dx} = 20x^3$
Standard formula $y = ax^n$	$\frac{dy}{dx} = nax^{n-1}$

You will notice that the power becomes a multiplier, and that the power is decreased by 1.

Integration – this is the opposite of differentiation. It is often used to find the area underneath a graph's curve between points. This is useful for calculating work done and distance travelled by a car or truck (see Figure 6.12).

Compound interest – if money is invested in a bank, or other such institution, and at the end of each year the interest is added to the investment, the interest on the interest will gain interest. This is called compound interest.

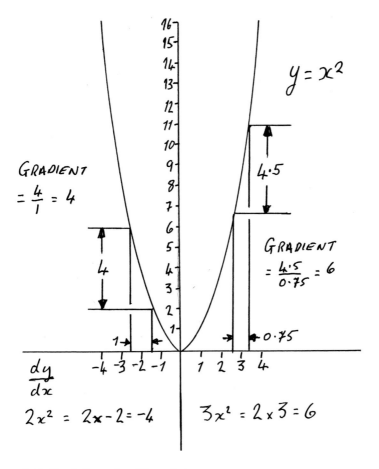

$$y = x^2$$

GRADIENT

$$= \frac{4}{1} = 4$$

4·5

GRADIENT

$$= \frac{4\cdot5}{0\cdot75} = 6$$

4

0·75

1

-4 -3 -2 -1   1 2 3 4

$\dfrac{dy}{dx}$

$2x^2 = 2x - 2 = -4$ $3x^2 = 2 \times 3 = 6$

Figure 6.11 Graph illustrating differentiation.

If a sum of £P, is invested at r% for n years. The initial value is £P, and after 1 year the value is:

$$£\left(P + \frac{Pr}{100}\right) \quad \text{or} \quad £P\left(1 + \frac{r}{100}\right)$$

$$= £\left[P\left(1 + \frac{r}{100}\right) + P\left(1 + \frac{r}{100}\right) \times \frac{r}{100}\right]$$

$$= £\left[P\left(1 + \frac{r}{100}\right)\left(1 + \frac{r}{100}\right)\right]$$

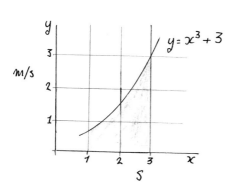

$$\int_1^3 y\,dx$$

$$\int_1^3 (x^3 + 3)\,dx$$

$$= \left[\frac{x^4}{4} + 3x\right]_1^3$$

$$= \left(\frac{3^4}{4} + 3(3)\right) - \left(\frac{1^4}{4} \times 3(1)\right)$$

$$= 29.25 - 3.25$$

$$= \underline{26\ metres\ squared}$$

Figure 6.12 Example of integration.

$$= £P\left(1 + \frac{r}{100}\right)^2$$

We derive the formula that the amount after n years using geometric progression whose first term is £P and whose common ratio is $\left(1 + \frac{r}{100}\right)$ is:

$$£P\left(1 + \frac{r}{100}\right)^n$$

For example, the value of £2,500 invested at 5% compound interest over 8 years:

$$P = 2500$$

$r = 5$

$n = 8$

$$£P\left(1+\frac{r}{100}\right)^{n}$$

$$2500\left(1+\frac{5}{100}\right)^{8}$$

$2500 \,(1.08)^{8}$

£3,693.64

Skills and Questions

Your college or training centre will have past-papers for mathematics examinations from previous years, you are advised to work through them. When solving the problems you should write the answers out fully, practise using numbers as much as you can so that you can answer the questions fully, neatly and quickly.

Chapter 7

Engineering Science

SI System and Common Units

SI stands for Système International, a system of measurement units which was developed following World War 2. There are several different systems of measurement in use throughout the world; but for the examinations with UK-based examining bodies SI only is used. It is worth noting that in countries such as Germany and Japan they use a version of SI; but with amendments and modifications. The Germans use DIN – Deutsch (Germany) Industrial Norm. The Japanese use Japanese Industrial Standards – JIS. The Americans use ANSI – American National Standards Institute – as well as SI. We'll also discuss some of the others as they are used in America too, for instance the imperial system – so called after the British Imperial Empire of the Victorian era which was used in the UK up until the 1970s – it is still used for many items around the world.

Imperial system of measurement for length uses inches, feet, yards and miles. For mass it uses ounces, pounds and tons.

Table 7.2 is an **approximate** guide.

Please note that the Glossary and List of Abbreviations sections of this book give more information about units and related terminology.

Decimals and Zeros

It's very easy to make mistakes with decimal calculations and the use of zeros. As you will have seen from Table 7.1 the standard or basic units are often too big or too small in value. So, there is a series of multiples and submultiples which are used to make calculations easier. For instance, kilo – meaning thousand – is added to metre giving kilometre. In other words, 1,000 metres. Going in the other direction we use milli – meaning

DOI: 10.1201/9781003284833-7

Table 7.1 SI Units

Quantity	Quantity Symbol	Unit	Unit Symbol
Length	l	metre	m
Mass	m	kilogram	kg
Time	t	second	s
Electric Current	I	ampere	A
Temperature	T	kelvin	K

Table 7.2 Imperial/SI Approximations

Quantity	Imperial	SI
Length	1 inch (in.)	25 mm
Length	1 foot (ft.)	300 mm
Length	1 yard (yd.)	900 mm
Length	39 inches	1 metre
Length	1 mile	1.6 kilometres
Mass	1 ounce	25 grams
Mass	1 pound	454 grams
Mass	2.2 pounds	1 kilogram (kg)
Mass	1 ton	1000 kilograms

Table 7.3 Multiples and Sub-Multiples

Prefix	Symbol	Power	Number
giga	G	10^9	1,000,000,000
mega	M	10^6	1,000,000
kilo	k	10^3	1000
hecto	h	10^2	100
deca	da	10^1	10
deci	d	10^{-1}	0.1
centi	c	10^{-2}	0.01
milli	m	10^{-3}	0.001
micro	μ	10^{-6}	0.000,001

one thousandth – when we are talking about the very low voltages in vehicle electronics – we say millivolts.

Scalar and vector – when you are throwing a ball it requires force. If it is a cricket ball and you want to knock the bails off it will need sufficient force and to be in the direction of the wicket. That is, it needs sufficient

Table 7.4 Scalars and Vectors

Scalars	Vectors
Distance	Displacement
Speed	Velocity
Mass	Weight
Pressure	Force
Energy	Momentum
Temperature	Acceleration
Volume	Electric current
Density	Torque

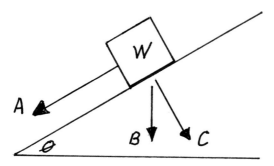

Figure 7.1 Vectors of box on slope.

force – referred to as magnitude and direction. This is called a vector. If you were just throwing the ball, not particularly at the wicket, it would be scalar. Vectors are usually represented on diagrams by arrows. The direction of the arrow represents the direction of the vector, the size of the arrow indicates the magnitude of the vector.

Vector Forces of a Box on a Slope

If we have a box on a slope there will be three vector forces. Force A which is parallel to the slope, force B which is the vertical and remains constant and force C which is at a right angle to the slope. The force B remains constant, that is if it weighs 100 N (mass × gravity), it will continue to weigh 100 N. That is, it is 100 N acting vertically, we sometimes say it is a perpendicular force. It is the weight of the box in the equation.

If we alter the angle of the slope, Θ (theta) the values for force A and force C will change. The forces obey trig functions. If we change the angle between 0° and 90° we can see the changes Table 7.5.

Table 7.5 Changes in Vector Forces

θ degrees	Force A − Wsin	Force C − WcosΘ
0	0	100
10	17.3	98.4
20	34.2	93.6
30	50	86.6
40	64.2	76.6
50	76.6	64.2
60	86.6	50
70	93.6	34.2
80	98.4	17.3
90	100	0

Accurate Measuring

Accurate measuring – called metrology in engineering terminology – is used extensively in manufacture. Five pieces of equipment which are used in manufacture are:

Co-ordinate measuring machine – CMM –When we are making a bracket, or pivot point to fit a mechanical piece of equipment, the first important measurements will be the mounting points. All other measurements will be taken from these points which we call co-ordinates – like the points on a map. A CMM measures distances from co-ordinates, even on the most irregular-shaped object such as a vehicle body. It does this to an accuracy of one micron (1 µm, a millionth of a metre). This is necessary as often the first prototypes are made by hand from the contours of clay models – called a buck. Therefore, the exact measurements may not be known.

Granite block – this is what its name says, a gigantic block of granite on which an engineered item, such as a car, can be stood. This block of granite will have a mass greater than that of the vehicle and will be supported on many hydraulic columns so that it can be kept perfectly level. Even the most level road surface will have a natural curvature – the curvature of the Earth. This granite block is made dead square. It enables dead accurate measurements to be taken of the subject being measured – measurements which can be used during the construction process. For the construction of items such as window frames, this is usually undertaken on jigs – sets of metal rails which support the frame members, with attachments to position the fittings and hangers. The measurements from the granite block are used to inform the measurements for the production jig. For quality sampling, complete items as big as cars can be measured on the granite blocks.

Lasers –can be used for positioning component mountings during construction. Having accurate measurements – using the CMM and position measurements from the granite block. It is possible to transfer this information to a jig to ensure that the components are correctly positioned before fixing takes place – either welding or bonding. This is done by setting up one or more lasers on the outer part of the jig that will shine a light, or several spots will converge together, when the part is accurately positioned. The part can then be bonded or welded into place most accurately. This is used widely in vehicle production.

Scanners – measuring an irregular-shaped object, such as a fuel tank, can be done without starting to mark the object, then the measurements may be transferable to metal to make another one. This has two main uses. In manufacturing, the scanner can be used to take the profile from the design mock up – or first hand-made metal part – and convert it into a CADCAM file for the manufacture of a press die. The other is to scan in the profile of existing parts to aid manufacturing new ones.

Wind tunnel – increasingly used to measure drag.

$$\text{drag force} = \tfrac{1}{2} \text{ air density} \times \text{velocity squared} \times \text{frontal area} \\ \times \text{drag co-efficient}$$

This might seem a lot of variables; but it isn't really. Most wind tunnels will give some form of read out of force, or you attach streamers to actually see what is happening. You can then alter the wind speed by turning the fan motor up. Altering the position of a component or fairing shape will change the frontal area and changing the shape, or material, will alter the drag co-efficient.

Capacity and Volume

Liquids in the UK and Europe are sold in litres, this may be in parts of a litre or multiples of litres – for example half-litre or maybe five litres. In the USA it may be sold in pints or gallons.

A litre is defined as the volume of 1000 cubic centimetres – 1000 cc. Water has a mass of 1 kg per litre at a temperature of 4°C. This is often referred to as density. Oil and other lubricants are lighter than water and have a density of about 0.9 kg per litre. A cubic centimetre is also known as a millilitre (ml).

Sometimes the term relative density (rd), or specific gravity (sg) is used. Both sets of words mean the same thing. That is the density of the oil is compared to the density of water. So, the oil – irrespective of its volume – if one litre weighs 0.9 kg will have a rd of 0.9.

Still used by some companies, and more so in the USA, are pints and gallons. Beer is also frequently sold in pints too. Be aware that English pints and gallons are different to American ones.

English pints are made up of 20 fluid ounces – a little obscure – that means the volume of 20 ounces of water at 17°C. An English pint of water weighs a pound and a quarter. An English gallon is 8 English pints weighing 10 pounds. It is defined as 4.54 litres.

An **American gallon** is defined as 231 cubic inches, that is 3.78 litres. In water it weighs 8.34 lb. So, it is considerably smaller. An American pint is 16 fluid ounces.

The volume changes with temperature – the volumes of both oil and beer are legally measured in the UK at 16°C.

Temperature and Heat

These two scientific terms are often mis-used, so let's get them cleared up so that you know what you are talking about when welding or brazing.

Temperature – this is the hotness or coldness of an object. There are three temperature scales in use:

Celsius (C) – also known as Centigrade because it has 100 degrees in it. It is related to the freezing and boiling point of water. Water freezes at 0 degree Celsius (0°C) and boils at 100°C.

Fahrenheit (F) – water freezes at 32°F and boils at 212°F.

Kelvin (k) – just uses the letter k (no degree symbol) – it is the absolute temperature scale. 0°C equals 273 k. 100°C equals 373 k. Absolute zero – the lowest temperature achievable – is 0 k which equals –273°C

Heat – this is the amount of energy used to raise the temperature of something. Heat is a form of energy. It takes 4200 joules of heat energy to raise the temperature of one kilogram of water 1 degree Celsius. Water is said to have a specific heat of 4200 J/kg C.

When you are using a gas torch to warm up something, for instance a piece of metal which you wish to bend, you will note that it takes time. Different metals take different times, and larger pieces take longer that smaller ones of the same metals. The longer time means that it is using more gas, this means more heat. As an example, typically 1 kg of propane will give 50 MJ.

Force and Pressure

These two terms are also often confused, or mis-used, let's clarify them – it'll come in useful when you are pushing and bending, or straightening something.

Force – we often use this in calculations; but don't necessarily understand it. The unit of force is the newton (N), named after Sir Isaac Newton who first discovered it. He noticed that if anything was dropped it would go to the ground. This is due to the force of gravity (G). The further an item drops the faster it goes – this is called acceleration due to gravity. The rate of acceleration on Earth is typically 9.81 m/s/s. For terms of simple calculations, we often use 10 m/s/s as the equivalent of G

Force (N) = Mass (kg) × Acceleration (G)

> **Tech note**
>
> Rate of acceleration of 9.81 m/s/s may also be written 9.81 ms^{-2} and velocity, such as 10 m/s may be written 10 ms^{-1} the use of indices makes calculations easier; but are not always clear to non-engineers, or non-technical people.

So, imagine an average person – mass 65 kg, stood on the pavement.

Force of person's feet on pavement = 65 kg × 10 (value of G)
= 650 N

Pressure – is force divided by the cross-sectional area

Pressure (N/m^2) = Force (N)/Cross-sectional area (m^2)

To avoid confusion with other units, the term Pascal (Pa) is used for 1 N/m^2 pressure.

The pressure of 1 Pascal (Pa) is very low, imagine an apple on your desk top – that's about 1 Pa. So, we tend to use the term bar – this is equivalent to normal barometric pressure. 1 bar equals 101.3 kPa. The air compressor which provides air pressure for your power tools generates about 10 bar in pressure. Vehicle tyres are typically inflated to between 1.5 and 2 bar.

Underwater pressure increases with depth. The deeper you go, the greater the pressure. The formula for this is:

Pressure = h ρ g

So the water pressure on a diving bell 2000 m under the surface of the ocean will be:

h = depth under water 2000 m

ρ (Greek rho) = density of sea water 1100 kgm^{-3}

g = gravity 9.81ms^{-2}

Pressure = 2000 × 1100 × 9.81

21.582 × 106 Pa or 215 bar

Tech note

Blaise Pascal lived in France from 1623 to 1662. His breakthrough in the science of pressure was discovering that the pressure in a liquid is equal in all directions – Pascal's law.

Amps, Volts, Ohms, Watts and Kirchhoff

Electricity is easy to understand if you get the basic terms clear. Although you can't see electricity it behaves in a similar way to water. Providing that the plumbing in your workshop is connected to the mains supply, when you turn the tap on water will flow out. Water comes out under pressure – pressure from the main supply. The consumer standard for water pressure in the UK is enough pressure to fill a 4.5-litre bucket in 30 seconds. Typically, about 2 bar.

When we talk about electricity the switch replaces the tap, the voltage (V) replaces the pressure and the amps (I) replaces the bucket full of water. So, the voltage needs to be high enough to force enough amps through to light up your lamp. Items in a circuit provide a resistance (R), like the tap, they can slow or stop the flow of electricity.

Safety note

ELECTRICITY CAN KILL YOU

Please be aware that any voltage or amperage of electricity can kill you. It can also give you a non-lethal shock which can make you flinch or jump and cause personal injury by hitting a rotating part, or a hot part.

Ohm's law – this is the relationship between amps (I) and volts (V) which will give a value for the resistance measured in ohms (R).

$I = V/R$

Watt (W) – this is a measure of power. If you look at light bulbs, they will usually have their power rating on them, the same applies to electric motors and heating elements. It is normal to state the power and the voltage on these electrical items. With LED lights the equivalent wattage is often given in two ways. For instance, some mains LED lights are rated as 9 W = 100

W. That is, they consume 9 W but give the equivalent light of a 100 W tungsten bulb – the older-type light bulb. They still work off a 230 V mains power supply.

Watts (W) = Volts (V) × Amps (I)

Tech note

I = current in Amps.
A is often used in a colloquial, or less formal way, also meaning Amps.

If you have the wattage and the voltage you can work out the current consumption in amps.

$I = W/V$

Knowing the current consumption is useful when you are fault finding, or wiring up a new component. Fault finding you can use an induction ammeter to see the actual current flowing. When fitting a component, it allows you to choose the correct cable size.

Kirchhoff – at any junction in an electrical circuit the current flowing into the junction will equal the current flowing out. This gives you more information when testing a circuit – finding where the current is flowing to.

Safety note

Always isolate the circuit concerned
Disconnect the battery where appropriate

Friction

Skidding – happens when the friction between the tyres and the road is not sufficient to keep a vehicle on a course. Friction is the ratio between the force acting downwards on the tyre (weight – W) and the force (F) needed to slide the tyre over the road. Not rolling the tyre; but making it skid. On a normal road with a good tyre, this ratio, μ (Greek letter mu) expressed as a decimal fraction, is about 0.8. If the road is covered in ice, or wet leaves, the ratio can be as low as 0.01 – in other word it can be pushed along with its brakes on.

μ = Force/Weight

Some Common Laws of Mechanical Engineering

Newton's laws are about force and acceleration, they are numbered:

First law – a body – a piece of metal, for instance – will either stay where it is, or continue moving uniformly unless another force is applied to it.

Second law – the force on a body is equal to its mass multiplied by its rate of acceleration. Usually expressed in the form Force (F) = Mass (m) × Acceleration (a)

Third law – when a force is applied to a second body the second body will be exerting a force backwards. Sometimes said as: to each action there is an equal and opposite reaction.

Hooke's law – is about how metal reacts to force. Metal stretches by an amount (X) proportionally to the force (F) applied to it until it gets to its elastic limit when it breaks. The amount of stretch depends of the type of metal, the metal type is given a constant (k)

$$F = k\,X$$

Impact and Momentum

Momentum – when a vehicle is travelling along a road it possesses momentum – Newton's first law. That momentum (p) is the product of its mass (m) and velocity (v) – that is speed combined with direction.

$$p = mv$$

The heavier the vehicle and the faster the speed, the greater the momentum. This is why we have speed limits.

Kinetic energy – we know that the moving vehicle possesses momentum and again applying Newton's laws – this time the second law, we have to apply a force to stop.

Work Done and Power

Work done (sometimes just called work) is the force applied multiplied by the distance travelled. The unit is the joule (J) named after the 19th-century Lancashire scientist James Prescott Joule.

Work done = force × distance

joule = Newton × metre

Power is work done per unit time, that is how much work can be done in a given time period. Power is expressed in Watts named after the 18th-century Scottish engineer James Watt.

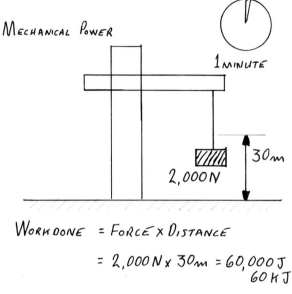

MECHANICAL POWER

1 MINUTE

30m

2,000N

WORK DONE = FORCE × DISTANCE

= 2,000N × 30m = 60,000 J
                   60 kJ

POWER   = $\dfrac{WORK\ DONE}{TIME}$

          = $\dfrac{60,000\ J}{60s}$ = 1,000W = 1kW

Figure 7.2 Mechanical power.

$$Power = \frac{Work\,done}{Unit\,time}$$

Work done is given in joules, time in seconds.

$$Watt = \frac{J}{s}$$

Typical question

A bus is travelling at 50 kph (13 ms^{-1}), it has a mass of 8000 kg.

a) What is the force needed to stop it in 100 m?
b) What happens to the energy in braking?

ELECTRICAL POWER

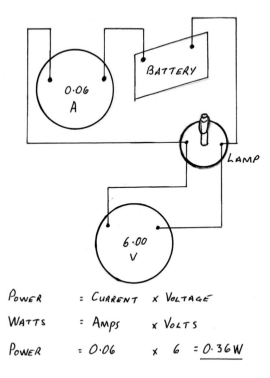

POWER   = CURRENT   x VOLTAGE

WATTS   = AMPS   x VOLTS

POWER   = 0.06   x 6 = 0.36 W

Figure 7.3 Electrical power.

Answers:
a)

KE = ½ m v²

½ 8000 × (13 ms⁻¹)²

67600 J (676 × 10³ J)

Work done (J) = force × distance

67600 J = force × 100 m

Therefore, Force = 67600/100 = 6760 N

b) The brakes convert the kinetic energy into heat energy which is dissipated into the atmosphere.

Gravitational Potential and Kinetic Energy

If we place a glass, or any other object, on a shelf it will possess potential energy. The higher the shelf the greater the amount of potential energy. So, if the glass falls off a higher shelf it is more likely to break that than if it fell off a lower one.

Newton's second law states:

Force = mass × acceleration

F = m a

a = gravitational acceleration when we consider the glass falling off the shelf. So, the longer it is falling – from the higher shelf – the greater will be the rate of acceleration.

Weight = mass × gravity

W = m g

So the heavier the glass, the more likely it is to break.

Gravitational potential energy = m g h

Or we may say:

Gravitational potential energy = W h

Hooke's Law and Young's Modulus

Robert Hooke was a 17th-century British polymath – that is, he was an expert in a number of fields of study covering a wide range of the arts and the sciences. Thomas Young was an 18th-century polymath – so wide was his range of studies that he is said to be the last man to know everything.

Hooke's law tells us that the amount by which a material stretches is proportional to the load applied until it reaches breaking point. This is useful for items like springs, we can calculate how much a spring will compress or stretch with the load applied. Young's modulus tells us about the overall strength of a material. Young worked as a surgeon in London and dealt with many of the injured soldiers of that time – Britain at that time was at war with both France and America – his interest was how much load a leg could take before being broken.

Experiment for Young's Modulus of Copper

1. Equipment: a piece of copper wire about 2m long, a metre rule, weights and hanger.
2. Calculate the cross-sectional area of the wire.

$$= \pi D^2/4$$

$$D = 0.3 \text{ mm}$$

$$= 3 \times 10^{-4} \text{ m}$$

$$D^2 = 9 \times 10^{-8}$$

$$= \pi \times (9 \times 10^{-8})$$

$$= 7.068 \times 10^{-8} \text{ m}^2$$

3. Set up the apparatus, add load in 1 N steps (100 g) noting the extension in metres for each extension in a table.
4. Plot a graph.
5. Calculate the results

Stress = force/area

Strain = extension/original length

Table 7.6 Load–Extension Table

No	Load – N	Original length – m	New length – m	Extension – m	Comment
1	1	1	1.000	0.000	No stretch
2	2	1	1.000	0.000	No stretch
3	3	1	1.001	0.001	
4	4	1	1.001	0.001	
5	5	1	1.002	0.002	
6	6	1	1.002	0.002	
7	7	1	1.005	0.005	
8	8	1	1.008	0.008	
9	9	1	1.025	0.025	Big increase
10	10	1	1.050	0.050	
11	11	1	1.080	0.080	
12	12	1	1.135	0.135	Can be seen
13	13	1	Wire broke		Broke

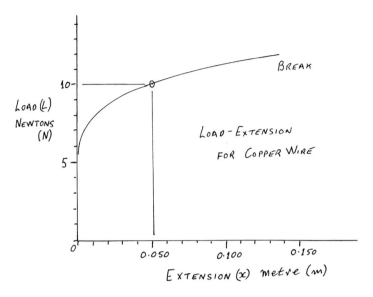

Figure 7.4 Load–extension graph.

Young's modulus (Y) = stress/strain

$$= \frac{\dfrac{Force}{Area}}{\dfrac{Extension}{original\,length}}$$

$$= \frac{Force \times length}{area \times extension}$$

Taking a mid-point on graph where 10N gives 0.050 m extension

$$= \frac{10 \times 1}{(7.068 \times 10 - 8) \times 0.050} \text{ should read } 10^{-8}$$

$$= 10/3.534 \times 10^{-9}$$

$$= \underline{2.82 \times 10^2 \text{ Pa}}$$

Boyle's and Charles' Laws

Robert Boyle was a 17th-century chemist working in the family brewery. Beer was drunk by most people as it was, in effect, clean water, it had less than one percent alcohol. He discovered that there was a relationship between pressure and volume. About 100 years later in the 18th-century, Jacques Charles, a balloonist, discovered that his balloon deflated when the temperature went down. Charles discovered the relationship between temperature and volume. The laws were combined to develop the combined gas laws.

The gas laws for a fixed mass of gas are:

P = pressure

V = volume

T = temperature

C = a constant – depending on the gas being measured.

Boyle's law

$PV = C$

Charles' law

$V/T = C$

Pressure law

$P/T = C$

These lead to the **ideal gas equation**

$PV/T = C$

A more useful way of writing the ideal gas equation, for solving problems, is to consider that the gas will retain the relationship between pressure, volume and temperature in the same proportions when any one of the variables is changed. This is simply expressed as:

$P_1 V_1/T_1 = P_2 V_2/T_2$ and so on

Internal combustion engine – petrol engine, this is a good example of the use of the gas laws. On the induction stroke, inlet valve open and piston travelling down, air and petrol are drawn into the cylinder at ambient

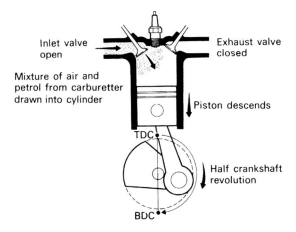

Inlet valve
open

Exhaust valve
closed

Mixture of air and
petrol from carburetter
drawn into cylinder

Piston descends

TDC

Half crankshaft
revolution

BDC

a *Induction stroke*

Figure 7.5 Petrol engine induction stroke.

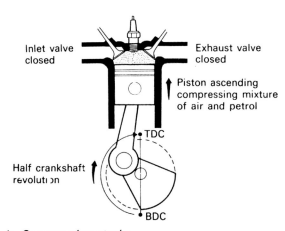

Inlet valve
closed

Exhaust valve
closed

Piston ascending
compressing mixture
of air and petrol

TDC

Half crankshaft
revoluti on

BDC

b *Compression stroke*

Figure 7.6 Petrol engine compression stroke.

temperature, about 20°C. The inlet valve closes as the piston ascends up the bore compressing the mixture of petrol and air into the combustion chamber. The compression ratio on a typical car engine is about 9:1. The temperature and pressure in the combustion chamber have increased. At this point a high-tension electrical spark ignites the petrol–air mixture and causes rapid combustion. The temperature increases to about 2000°C and

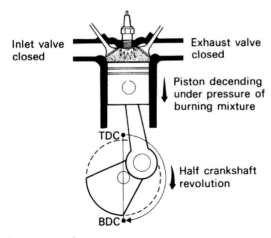

Inlet valve closed

Exhaust valve closed

Piston decending under pressure of burning mixture

TDC

Half crankshaft revolution

BDC

c *Power stroke*

Figure 7.7 Petrol engine power stroke.

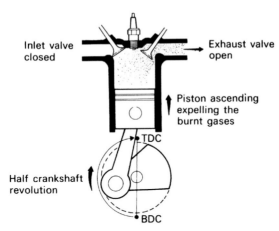

Inlet valve closed

Exhaust valve open

Piston ascending expelling the burnt gases

TDC

Half crankshaft revolution

BDC

d *Exhaust stroke*

Figure 7.8 Petrol engine exhaust stroke.

the pressure to about 70,000 kPa. This pressure pushes the piston down the bore to turn the crankshaft and hence move the vehicle.

Skills and Questions

This chapter has set out a number of scientific concepts and principles. If you can carry out experiments related to any of them at college, you ought to take that opportunity, especially the box on an inclined plane, and Hooke's law and Young's modulus. When you are carrying out tasks in the workshop take notes of readings and connections.

Chapter 8

Engineering Materials

Engineers tend to classify materials into two major groups, each with two sub-groups. The major groups are metallic materials and non-metallic materials. We'll look at each in turn (Table 8.1).

Metallic Materials

The metallic group is divided into two sub-groups, these are **ferrous metals** and **non-ferrous metals**. Ferrous simply means iron, all ferrous metals contain iron. Non-ferrous metals do not contain iron.

Iron is dug from the ground and heated in a furnace – there are several different types of furnaces – and mixed with carbon to form steel. Steel has been used for engineering and construction since about 500 BC. When we talk about steel it is important to realise that there are several major categories of steel: low-carbon, medium-carbon, high-carbon and many types of alloy steel. When we talk about alloy steel, we simply mean that it is steel mixed with another element.

Tech note

Ferrous – contains iron.
Alloy – a mixture of a metal and another element.

Manufacture of Steel

Iron ore which is dug up from the ground is fed into a blast furnace together with limestone and coke. The coke is used as a source of heat and the limestone as a flux, that is an agent which cleans and helps the flow of the metal. It separates the metal from the impurities in the mixture. The molten metal is then poured out of the furnace into moulds to form what are called

DOI: 10.1201/9781003284833-8

Table 8.1 Metallic Materials and Non-Metallic Materials

Engineering Materials

Metallic Materials

Ferrous – contains iron	Non-ferrous – does not contain iron
Iron in various forms	Aluminium
Low-carbon steel	Brass – copper and zinc
Medium-carbon steel	Bronze – copper and tin
High-carbon steel	Chromium
Alloy steel	Copper
	Titanium

Non-Metallic Materials

Natural – occur in nature	Synthetic – man-made materials
Leather	Carbon fibre
Wood	GRP – glass fibre
Wool	Vegan leather
Bamboo	Thermo-plastics
Cotton cloth	Thermo-setting plastics

pigs – chunks of iron which resemble the shape of a pig's body. Because of the burning process the pig iron contains between 3–4% carbon.

The pig iron is then changed into steel by re-heating in a furnace and blasting with air to reduce the carbon content to between 0.08–0.20%. The term blast furnace is used, though there are other types of processes.

Casting

The steel is cast into ingots, or into a continuous rolling slab depending on the process and purpose of the material.

Pickling

The next stage is to remove the black scale from the surface of the metal, this is a process called pickling – the steel is run through a bath, or shower, of either hydrochloric acid or sulphuric acid. This ensures that the surface of the steel is clean.

Cold Rolling and Hot Rolling

To make sheet steel, which may be converted into tubing by bending and joining, the ingots, or continuous slab may be rolled either when it is hot or when it is cold. Rolling changes the structure of the steel and needs to be followed by annealing and tempering stages. Tubing made from sheet steel is used in the construction of some utility bicycles, and the conduit tubing and trunking used in building construction which carries only minimum stresses.

Annealing and Tempering

Annealing is a method of treating the steel, taking away the internal stress. It needs to be stress free, or softened, to be able to be worked into the desired shape. Tempering is a process of making it the correct hardness. The annealing process usually means heating and cooling the steel in a controlled, oxygen-free atmosphere. The tempering involves re-heating to a set temperature and cooling at a specific rate. As you can see, this process uses a large amount of energy – the material has been heated and cooled four times – original casting, blasting, annealing and tempering.

Hot Drawing and Cold Drawing

Sheet steel can be bent into tubes and seam welded. The problem is the seam is a weakness and may tear apart under stress. Therefore, tubing made in this way tends to be made thicker to give the required factor of safety.

Tech note

Factor of safety is the number of times that the maximum load that a component can carry is divided by the expected load, this is expressed as a ratio or percentage. If a bicycle tube can carry a load of 650 kg before breaking, and the load is 65 kg (weight of a typical rider), then the calculation is 650/65 = 10. The factor of safety is 10.

Reynolds in the UK and Columbus in Italy developed ways of drawing tubing without seams, seamless tubing. Being seamless the tubing is equally strong across its entirety.

The hot ingot is held in a die with a plug, the tube is formed by being pulled over the plug. This process may be carried out several times to get the required tube wall thickness. The tubes are usually made in lengths of 2.4 metres (8 ft).

Classifications of Steel

Steel is a highly developed product and is classified in a number of ways, general classifications are:

- Cold forming steels
- Carbon steels
- Alloy steels
- Free cutting steels
- Spring steels
- Rust-resisting and stainless steels

Cold forming steels – these are in effect sheet steel, used on pressed steel components. Components such as bicycle parts and cutlery are made from pressed steel as they can be made in very large numbers very quickly. Using automated presses, components can be made in less than five seconds – that is up to 18,000 per day. The process for the manufacture of a component may requires two stages. First the material is cut to size – that is, it is stamped out like you might cut a biscuit out of pastry. Then it is moved to a machine which will bend it, also called forming, into shape against a die. The cutters for the shape and the dies for the forming can be changed in these machines in minutes so that a factory can make air boxes in the morning and exhausts in the afternoon – or any other similar product. Such companies tend to supply pressed products to a range of different industries, such as electronics manufacturers, the automotive industry and the construction industry.

    Carbon steels – these are used for a large number of small components, such as gears, axles and bearings. There is a large variation in carbon steels, this leads to a set of general classifications of carbon steel:

- Low-carbon steel – also called mild steel – 0.10–0.25% carbon
- Medium-carbon steel – 0.20–0.50% carbon
- High-carbon steel – 0.50–2.00% carbon

Low-carbon steel is soft, ductile and malleable and therefore can be easily formed into shape. It cannot be hardened and tempered by heating and quenching; but it can be case-hardened and it will work harden. Case-hardening is a process of coating the surface of the steel component with a high-carbon content chemical and heating to a set temperature. When the component cools the surface is hard like high-carbon steel and the underside remains soft and malleable. This process is used on bearing surfaces, if you look at a bearing closely you will be able to see the different colours of the metal. The advantages of this are that the axle and hub can be made of low-carbon steel, which is both easier to machine and cheaper to buy and then given a wear-resistant surface for the bearing.

    Low-carbon steel is usually sufficiently strong for many metal components, it is also cheap and plentiful, most steel suppliers can offer this readily from stock.

Medium-carbon steel is much tougher and not as easy to bend or machine; it can be hardened and tempered.

Alloying Metals Used with Steel

Chrome – a lustrous, brittle hard metal used to add corrosion resistance. It is the main additive in stainless steel. Abbreviation is **Cr**.

Manganese – used in stainless steel to resist corrosion. Increases hardenability and tensile strength. Abbreviation is **Mn**.

Molybdenum – used to enhance strength, improve the hardenability and weldability properties and add toughness. It also improves corrosion resistance and high-temperature deformation. Abbreviation is **Mo**.

Vanadium – gives added resistance to corrosion and resistance to acids and alkalis. Abbreviation is **V**.

Metal Forming Processes

Extrusion – shaping components by forcing the metal through a shaped hole. One way to explain this is to think about piping icing on to a cake. When you squeeze the icing bag it comes out through a shaped end with a profile.

Hydroformed – malleable metals such as aluminium can be formed into fairly complex shapes by hydroforming. This practice is extensively used on large-diameter shaped frame tubes, such as those typically used on MTB bicycles. The basic tube is put into a die assembly and then a liquid, either water-based or oil-based, is fed under pressure into the tube, forcing it outwards into the shape of the die.

Mar-aging steel, also written **maraging steel** without the hyphen. This word is a combination of martensitic and aging. It is a process of adding toughness and strength to low-carbon ultra-high strength steels by heating to a high temperature for several hours before cooling. Ultra-high-strength steels get their strength from intermetallic compounds, not added carbon. The compounds may include: cobalt, molybdenum, titanium and niobium.

Stainless precipitation hardening steel – these are low-carbon steels which have fairly high percentages of manganese, chromium, nickel, copper and titanium. The non-ferrous metals precipitate to make the steel hard. Precipitation means falling, a word that weather forecasters use for raining. In this case it is the even distribution of these compounds of non-ferrous metals in the steel, like rain drops, that make the steel hard.

Cold worked – means steel which is rolled out when it is cold. This changes the grain structure and so makes the metal harder and stronger; but reduces ductility.

Seamless – tubing made from billet; not rolled and seam welded.

Air-hardening steel – this is fairly high carbon, 0.5–2%, with the addition of molybdenum, chromium and manganese. It is hardened by heating to between about 800 and 900°C then cooling in air. The heating may be carried out in a vacuum furnace.

Butting – making the tubes thicker in places. Double butting is the most common, the tubes are butted at each end where they join the other tubes. This is typically used in bicycle and motorcycle frames.

Aluminium

Tech note

You must keep aluminium completely separate from steel and other metals to prevent **cross-contamination.** This means completely separate storage facilities and keeping all the tools separate too.

Production of Aluminium

Aluminium has been known about for about for 150 years. This may seem a long time; but in terms of metal working means it is relatively new, the industrial history of aluminium did not begin until 1886 when Paul Heroult in France discovered the basis of the present-day method of producing aluminium. Aluminium is now produced in such quantities that in terms of volume it ranks second to steel among the industrial metals. Aluminium of commercial purity contains at least 99% aluminium, while higher grades contain 99.5–99.8% of pure metal.

In the production of aluminium, the ore bauxite is crushed and screened, then washed and pumped under pressure into tanks and filtered into rotating drums, which are then heated. This separates the aluminium oxide from the ore. In the next stage the aluminium oxide is reduced to the metal aluminium by means of an electrolytic reduction cell. This cell uses powdered cryolite and a very heavy current of electricity to reduce the aluminium oxide to liquid metal, which passes to the bottom of the cell and is tapped off into pigs of aluminium of about 225 kg each. In the UK this process is mainly carried out in the north of Scotland as hydro-electric power is readily available relatively cheaply.

Types of Aluminium Sheet

Sheet, strip and circle blanks are sold in hard and soft tempers possessing different degrees of ductility and tensile strength. Sheet is supplied in gauges down to 0.3 mm, but it is generally more economical to order strip for gauges less than 1.6 mm.

Manufacturing Process

Sheet products are first cast by the semi-continuous casting process, then scalped to remove surface roughness and preheated in readiness for hot rolling. They are first reduced to the thickness of plate, and then to sheet if this is required. Hot rolling is followed by cold rolling, which imparts finish and temper in bringing the metal to the gauge required. Material is supplied in the annealed, that is soft condition, and in at least three degrees of hardness, H1, H2 and H3 (in ascending order of hardness).

Aluminium Alloys

From the reduction centre the pigs of aluminium are re-melted and cast into ingots of commercial purity. Aluminium alloys are made by adding specified amounts of alloying elements to molten aluminium. Some alloys, such as magnesium and zinc, can be added directly to the melt, but higher-melting-point elements such as copper and manganese have to be introduced in stages. Aluminium and aluminium alloys are produced for industry in two broad groups:

1   Materials suitable for casting
2   Materials for the further mechanical production of plate sheet and strip, bars, tubes and extruded sections.

In addition, both cast and wrought materials can be subdivided according to the method by which their mechanical properties are improved:

1.   **Non-heat-treatable alloys** – Wrought alloys, including pure aluminium, gain in strength by cold working such as rolling, pressing, beating and any similar type of process.
2.   **Heat-treatable alloys** – these are strengthened by controlled heating and cooling followed by ageing at either room temperature, or at 100–200°C.

The most commonly used elements in aluminium alloys are copper, manganese, silicon, magnesium and zinc. The manufacturers can supply these materials in a variety of conditions.

Rubber

There are two types of rubber, natural rubber made from plant latex – known commonly as rubber trees, these grow mainly in Malaysia and Indonesia. The other is synthetic, made from petro-chemical materials.

The value of rubber lies in the fact that it can be readily moulded or extruded to any desired shape, and its elastic quality makes it capable of filling unavoidable and irregular gaps and clearances. Some classifications of rubber include:

1. Moulded latex foam
2. Low-grade fabricated polyether
3. Fabricated polyether
4. Moulded polyether
5. Fabricated polyester
6. Polyvinyl chloride foam
7. Reconstituted polyether.

Plastics

The word plastic really means that something can be bent and will stay in that shape. The opposite is elastic, which is something that can be bent and will spring back to shape. Think of plasticine modelling material and elastic bands to get the idea.

Thermoplastics and thermosetting plastics

The simplest way of classifying plastics is by their reaction to heat. This gives a ready subdivision into two basic groups: thermoplastics and thermosetting plastics. Thermoplastic materials soften to become plastic when heated, no chemical change taking place during this process. When cooled they again become hard and will assume any shape into which they were moulded when soft. Thermosetting materials, as the name implies, will soften only once. During heating a chemical change takes place and the material cures; thereafter the only effect of heating is to char or burn the material. As far as performance is concerned, these plastics can be divided into three groups, as described below.

General-Purpose Thermoplastics

- Polyethylene
- Polypropylene
- Polystyrene
- SAN (styrene/acrylonitrile copolymer)

- Impact polystyrene
- ABS (acrylonitrile butadiene styrene)
- Polyvinyl chloride (PVC)
- Poly vinylidene chloride
- Poly methyl methacrylate
- Poly ethylene terephthalate

Engineering Thermoplastics

- Polyesters (thermoplastic)
- Polyamides
- Polyacetals
- Polyphenylene sulphide
- Polycarbonates
- Polysulphone
- Modified polyphenylene ether
- Polyimides
- Cellulosics
- RIM/polyurethane
- Polyurethane foam

Thermosetting Plastics

- Phenolic
- Epoxy resins
- Unsaturated polyesters
- Alkyd resins
- Diallyl phthalate
- Amino resins

Plastics are used extensively for non-load-bearing components, covers and cowlings. They are also used for relatively low-load components, and components where there is little heat, and or low impact stress. Offering greater strength, and the same low weight are composite materials.

Reinforced Composite Materials

Introduction

Composite materials came into use in the automotive and boat building industries in the 1950s to fulfil the needs of small post-war car manufacturers, as metal was rationed for 12 years after the end of World War 2 small companies sought alternatives. The automotive market was developing at the same time as the small boat market; both developed along the same lines using a glass fibre and a resin lay-up procedure called glass reinforced plastics (**GRPs**) which is still used today by many kit car manufacturers and makers of boats and some motorcycle parts. The use of **composites** – a product made up of more than one material which is bonded together to provide special properties – became more specialised as **carbon-based technical materials** became available. Carbon fibre, as it is called, was originally invented and patented by The Royal Aircraft Establishment at Farnborough, UK. Bicycles started to be made from it in the 1990s as the materials became more readily available and methods of joining tubes and laying up became known. The usage of carbon fibre for motorcycles has until recently been limited to small parts, such as huggers. Now manufacturers are starting to experiment with it as a frame material. Given the use of moulds and autoclave equipment, the manufacture of frames in carbon fibre should be cheaper and require less skill than the construction of frames from metal.

Depending on the materials used in composite construction, the following properties of composites may influence their choice:

- Components can be produced on a one-off basis with minimum tooling
- Compound curvature can be produced with constant material thickness
- Extreme lightness for a given strength
- Resistant to corrosion
- Different finishes are available
- There is a style cache in the use of carbon fibre.

Basic Principles of Reinforced Composite Materials

The basic principle involved in reinforced plastic production is the combination of polyester resin and reinforcing fibres to form a solid structure. Glass-reinforced plastics are essentially a family of structural materials which utilise a very wide range of thermoplastic and thermosetting resins. The incorporation of glass fibres in the resins changes them from relatively low-strength, brittle materials into strong and resilient structural materials. In many ways, glass fibre-reinforced plastic can be compared to concrete, with the glass fibres performing the same function as the steel reinforcement and the resin matrix acting as the concrete. Glass fibres have high strength and high modulus, and the resin has low strength and low modulus. Despite this, the resin has the important task of transferring the stress from fibre to fibre, so enabling the glass fibre to develop its full strength.

Polyester resins are supplied as viscous liquids which solidify when the actuating agents, in the form of a catalyst and accelerator, are added. The proportions of this mixture, together with the existing workshop conditions, dictate whether it is cured at room temperature or at higher temperatures and also the length of time needed for curing. In common practice, pre-accelerated resins are used, requiring only the addition of a catalyst to affect the cure at room temperature. Glass reinforcements are supplied in a number of forms, including chopped strand mats, needled mats, bi-directional materials such as woven rovings and glass fabrics, and rovings which are used for chopping into random lengths or as high-strength directional reinforcement. Other materials needed are the releasing agent, filler and pigment concentrates for the colouring of glass fibre-reinforced plastic.

Among the methods of production, the most used method is that of contact moulding, or the wet laying-up technique as it is sometimes called. The mould itself can be made of any material which will remain rigid during the application of the resin and glass fibre, which will not be attacked by the chemicals involved, and which will also allow easy removal after the resin has set hard. Those in common use are wood, plaster, sheet metal and glass fibre itself, or a combination of these materials. The quality of the surface of the completed moulding will depend entirely upon the surface finish of the mould from which it is made. When the mould is ready the releasing agent is applied, followed by a thin coat of resin to form a gel coat. To this a fine surfacing tissue of fibre glass is often applied. Further resin is applied, usually by brush, and carefully cut-out pieces of mat or woven cloth are laid in position. The use of split washer rollers removes the air and compresses the glass fibres into the resin. Layers of resin and glass fibres are added until the required thickness is achieved. Curing takes place at room temperature but heat can be applied to speed up the curing time. Once the catalyst has caused the resin to set hard, the moulding can be taken from the mould.

Manufacture of Reinforced Composite Materials

When glass is drawn into fine filaments its strength greatly increases over that of bulk glass. Glass fibre is one of the strongest of all materials. The ultimate tensile strength (UTS) of a single glass filament (diameter 9–15 micrometres) is about 3,447,000 kN/m². It is made from readily available raw materials, and is non-combustible and chemically resistant. Glass fibre is therefore the ideal reinforcing material for plastics. In Great Britain, the type of glass which is principally used for glass fibre manufacture is E glass, which contains less than 1% alkali borosilicate glass. E glass is essential for electrical applications and it is desirable to use this material where good weathering and water-resistance properties are required. Therefore it is greatly used in the manufacture of composite body shells.

Basically, the glass is manufactured from sand or silica and the process by which it is made proceeds through the following stages:

1   Initially the raw materials, including sand, china clay and limestone, are mixed together as powders in the desired proportions.

2   The 'glass powder', or frit as it is termed, is then fed into a continuous melt furnace or tank.

3   The molten glass flows out of the furnace through a forehearth to a series of fiberising units usually referred to as bushings, each containing several hundreds of fine holes. As the glass flows out of the bushings under gravity it is attenuated at high speed.

After fiberising, the filaments are coated with a chemical treatment usually referred to as a forming size. The filaments are then drawn together to form a strand which is wound on a removable sleeve on a high-speed winding head. The basic packages are usually referred to as cakes and form the basic glass fibre which, after drying, is processed into the various reinforcement products. Most reinforcement materials are manufactured from continuous filaments ranging in fibre diameters from 5 to 13 micrometres. The fibres are made into strands by the use of size. In the case of strands which are subsequently twisted into weaving yarns, the size lubricates the filaments as well as acting as an adhesive. These textile sizes are generally removed by heat or solvents and replaced by a chemical finish before being used with polyester resins. For strands which are not processed into yarns it is usual to apply sizes which are compatible with moulding resins.

Glass reinforcements are supplied in a number of forms, including chopped strand mats, needled mats, bi-directional materials such as woven rovings and glass fabrics, and rovings which are used for chopping into random lengths or as high-strength directional reinforcements.

Types of Reinforcing Material

Woven Fabrics

Glass fibre fabrics are available in a wide range of weaves and weights. Lightweight fabrics produce laminates with higher tensile strength and modulus than heavy fabrics of a similar weave. The type of weave will also influence the strength (due, in part, to the amount of crimp in the fabric), and usually satin weave fabrics, which have little crimp, give stronger laminates than plain weaves which have a higher crimp. Satin weaves also drape more easily and are quicker to impregnate. Besides fabrics made from twisted yarns, it is now the practice to use woven fabrics manufactured from rovings. These fabrics are cheaper to produce and can be much heavier in weight.

Chopped Strand Mat

Chopped strand glass mat (**CSM**) is the most widely used form of reinforcement. It is suitable for moulding the most complex forms. The strength of laminates made from chopped strand mat is less than that with woven fabrics, since the glass content which can be achieved is considerably lower. The laminates have similar strengths in all directions because the fibres are random in orientation. Chopped strand mat consists of randomly distributed strands of glass about 50 mm long which are bonded together with a variety of adhesives. The type of binder or adhesive will produce differing moulding characteristics and will tend to make one mat more suitable than another for specific applications.

Needle Mat

This is mechanically bound together and the need for an adhesive binder is eliminated. This mat has a high resin pick-up owing to its bulk, and cannot be used satisfactorily in moulding methods where no pressure is applied. It is used for press moulding and various low-pressure techniques such as pressure injection, vacuum and pressure bag.

Rovings

These are formed by grouping untwisted strands together and winding them on a 'cheese'. They are used for chopping applications to replace mats either in contact moulding (spray-up), or translucent sheet manufacture of press moulding (pre-form). Special grades of roving are available for each of these different chopping applications. Rovings are also used for weaving, for filament winding and for pultrusion processes. Special forms are available to suit these processes.

Chopped Strands

These consist of rovings prechopped into strands of 6 mm, 13 mm, 25 mm or 50 mm lengths. This material is used for dough moulding compounds, and in casting resins to prevent cracking.

Staple Fibres

These are occasionally used to improve the finish of mouldings. Two types are normally available, a compact form for contact moulding and a soft bulky form for press moulding. These materials are frequently used to reinforce gel coats. The weathering properties of translucent sheeting are considerably improved by the use of surfacing tissue.

Application of These Materials

Probably chopped strand mat is most commonly used for the average moulding. It is available in several different thicknesses and specified by weight: 300, 450 and 600 g/m². The 450 g/m² is the most frequently used, and is often supplemented with the 300 g/m². The 600 g/m² density is rather too bulky for many purposes, and may not drape as easily, although all forms become very pliable when wetted with the resin. The woven glass fibre cloths are generally of two kinds, made from continuous filaments or from staple fibres. Obviously, most fabricators use the woven variety of glass fibre for those structures that are going to be the most highly stressed. For example, a moulded glass fibre seat pan and squab unit in a human-powered vehicle (HPV) would be made with woven material as reinforcement, but a detachable hard top for a trailer body would more probably be made with chopped strand mat as a basis. However, it is quite customary to combine cloth and mat, not only to obtain adequate thickness, but because if the sandwich of resin, mat and cloth is arranged so that the mat is nearest to the surface of the final product, the appearance will be better.

The top layer of resin is comparatively thin, and the weave of cloth can show up underneath it, especially if some areas have to be buffed subsequently. Chopped fibres do not show up so prominently, but some fabricators compromise by using the thinnest possible cloth (surfacing tissue as it is known) nearest the surface, on top of the chopped strand mat. When moulding, these orders are of course reversed, the tissue going on to the gel coat on the inside of the mould, followed by the mat and resin lay-up.

It is important to note that if glass cloths or woven mat are used, it is possible to lay up the materials so that the reinforcement is in the direction of the greatest stresses, thus giving extra strength to the entire fabrication. In plain weave cloths, each warp and weft thread passes over one yarn and under the next. In twill weaves, the weft yarns pass over one warp and under more than one warp yarn; in 2 × 1 twill, the weft yarns pass over one warp yarn and under two warp yarns. Satin weaves may be of multi-shaft types, when each warp and weft yarn goes under one and over several yarns. Unidirectional cloth is one in which the strength is higher in one direction than the other, and balanced cloth is a type with the warp and weft strength about equal. Although relatively expensive, the woven forms have many excellent qualities, including high dimensional stability, high tensile and impact strength, good heat, weather and chemical resistance, low moisture absorption, resistance to fire and good thermo-electrical properties. A number of different weaves and weights are available, and thickness may range from 0.05 mm to 9.14 mm, with weights from 30 g/m² to 1 kg/m², although the grades mostly used in the automotive field probably have weights of about 60 g/m² and will be of plain, twill or satin weave.

Carbon Fibre

This is another reinforcing material. Carbon fibres possess a very high modulus of elasticity, and have been used successfully in conjunction with epoxy resin to produce low-density composites possessing high strength.

Resins Used in Reinforced Composite Materials

The first man-made plastics were produced in the United Kingdom in 1862 by Alexander Parkes and were the forerunner of celluloid. Since then a large variety of plastics has been developed commercially, particularly in the last 25 years. They extend over a wide range of properties. Phenol formaldehyde is a hard thermoset material; polystyrene is a hard, brittle thermo-plastic; polythene and plasticised polyvinyl chloride (PVC) are soft, tough thermo-plastic materials; and so on. Plastics also exist in various physical forms. They can be bulk solid materials, rigid or flexible foams, or in the form of sheet or film. All plastics have one important common property. They are composed of macro-molecules, which are large chain-like molecules consisting of many simple repeating units. The chemist calls these molecular chains polymers. Not all polymers are used for making plastic mouldings. Man-made polymers are called synthetic resins until they have been moulded in some way, when they are called plastics.

Most synthetic resins are made from oil. The resin is an essential component of glass fibre-reinforced plastic. The most widely used is unsaturated polyester resin, which can be cured to a solid state either by catalyst and heat or by catalyst and accelerators at room temperature. The ability of polyester resin to cure at room temperature into a hard material is one of the main reasons for the growth of the reinforced plastics industry. It was this which led to development of the room temperature contact moulding methods which permit the production of extremely large integral units.

Tech note

Scientists are working on making carbon fibre materials bio-degradable.

Polyester resins are formulated by the reaction of organic acids and alcohols which produces a class of material called esters. When the acids are polybasic and the alcohols are polyhydric they can react to form a very complex ester which is generally known as polyester. These are usually called alkyds, and have long been important in surface coating formulations because of their toughness, chemical resistance and endurance. If the acid or alcohol used contains an unsaturated carbon bond, the polyester formed can react further with other unsaturated materials such as styrene or diallyl phthalate. The result of this reaction is to interconnect the different polyester units to form the three-dimensional cross-linked structure that is characteristic of thermosetting resins. The available polyesters are solutions of these alkyds in the cross-linking monomers. The curing of the resin is the reaction of the monomer and the alkyd to form the cross-linked structure. An unsaturated polyester resin is one which is capable of being cured from a liquid to a solid state when subjected to the right conditions. It is usually referred to as polyester.

Catalysts and Accelerators

In order to mould or laminate a polyester resin, the resin must be cured. This is the name given to the overall process of gelation and hardening, which is achieved either by the use of a catalyst and heating, or at normal room temperature by using a catalyst and an accelerator. Catalysts for polyester resins are usually organic peroxides. Pure catalysts are chemically unstable and liable to decompose with explosive violence. They are supplied, therefore, as a paste or liquid dispersion in a plasticiser, or as a powder in an inert filler. Many chemical compounds act as accelerators, making it possible for the resin-containing catalyst to be cured without the use of heat. Some accelerators have only limited or specific uses, such as quaternary ammonium compounds, vanadium, tin or zirconium salts. By far the most important of all accelerators are those based on a cobalt soap or those based on a tertiary amine. It is essential to choose the correct type of catalyst and accelerator, as well as to use the correct amount, if the optimum properties of the final cured resin or laminate are to be obtained.

Pre-Accelerated Resins

Many resins are supplied with an in-built accelerator system controlled to give the most suitable gelling and hardening characteristics for the fabricator. Pre-accelerated resins need only the addition of a catalyst to start the curing reaction at room temperature. Resins of this type are ideal for production runs under controlled work- shop conditions.

The cure of a polyester resin will begin as soon as a suitable catalyst is added. The speed of the reactions will depend on the resin and the activity of the catalyst. Without the addition of an accelerator, heat or ultraviolet radiation, the resin will take a considerable time to cure. In order to speed up this reaction at room temperature it is usual to add an accelerator. The quantity of accelerator added will control the time of gelation and the rate of hardening.

There are three distinct phases in the curing reaction:

Gel time This is the period from the addition of the accelerator to the setting of the resin to a soft gel.

Hardening time This is the time from the setting of the resin to the point when the resin is hard enough to allow the moulding or laminate to be withdrawn from the mould.

Maturing time This may be hours, several days or even weeks depending on the resin and curing system, and is the time taken for the moulding or laminate to acquire its full hardness and chemical resistance. The maturing process can be accelerated by post-curing. When the material is not fully matured it is referred to as being in its **green state**, or simply as **green**; this term is taken from the colour of the wood of a freshly chopped down tree.

Fillers and Pigments

Fillers are used in polyester resins to impart particular properties. They will give opacity to castings and laminates, produce dense gel coats, and impart specific mechanical, electrical and fire-resisting properties. A particular property may often be improved by the selection of a suitable filler. Powdered mineral fillers usually increase compressive strength; fibrous fillers improve tensile and impact strength. Moulding properties can also be modified by the use of fillers; for example, shrinkage of the moulding during cure can be considerably reduced. There is no doubt, also, that the wet lay-up process on vertical surfaces would be virtually impossible if thixotropic fillers were not available.

Polyester resins can be coloured to any shade by the addition of selected pigments and pigment pastes, the main requirement being to ensure thorough dispersion of colouring matter throughout the resin to avoid patchy mouldings.

Both pigments and fillers can increase the cure time of the resin by the dilution effect, and the adjusted catalyst and promoter are added to compensate.

Releasing Agents

Releasing agents used in the normal moulding processes may be either water-soluble film-forming compounds, or some type of wax compound. The choice of releasing agent depends on the size and complexity of the moulding and on the surface finish of the mould. Small mouldings of simple shape, taken from a suitable GRP mould, should require only a film of polyvinyl alcohol (PVAL) to be applied as a solution by cloth, sponge or spray. Some mouldings are likely to stick if only PVAL is used. PVAL is available as a solution in water or solvent, or as a concentrate which has to be diluted, and it may be in either coloured or colourless form.

Suitable wax emulsions are also available as a releasing agent. They are supplied as surface finishing pastes, liquid wax or wax polishes. The recommended method of application can vary depending upon the material to be finished. Hand application is with a pad of damp, good-quality mutton cloth or equivalent, in straight even strokes. Buff lightly to a shine with a clean, dry, good-quality mutton cloth. Machine at 1800 rpm using a G-mop foam finishing head. Soak this head in clean water before use and keep damp during compounding. Apply the wax to the surface. After compounding, remove residue and buff lightly to a shine with a clean, dry, good=quality mutton cloth.

Wax polishes should be applied in small quantities since they contain a high percentage of wax solids. Application with a pad of clean, soft cloth should be limited to an area of approximately 1 square metre. Polishing should be carried out immediately, before the wax is allowed to dry. This can be done either by hand or by machine with the aid of a wool mop polishing bonnet.

Frekote is a semi-permanent, multi-release, gloss finish, non-wax polymeric mould release system specially designed for high-gloss polyester mouldings. It will give a semi-permanent release interface when correctly applied to moulds from ambient up to 135°C. This multi-release interface chemically bonds to the mould's surface and forms on it a micro-thin layer of a chemically resistant coating. It does not build up on the mould and will give a high-gloss finish to all polyester resins, cultured marble and onyx. It can be used on moulds made from polyester, epoxy, metal or composite moulds. Care should be taken to avoid contact with the skin, and the wearing of suitable clothing, especially gloves, is highly recommended. These products must be used in a well-ventilated area.

Adhesives Used with GRP

Since polyester resin is highly adhesive, it is the logical choice for bonding most materials to GRP surfaces.

Suitable alternatives include the Sika technique, which is a heavy-duty, polyurethane-based joining compound. It cures to a flexible rubber which bonds firmly to wood, metal, glass and GRP. It is ideal for such jobs as bonding glass to GRP or bonding GRP and metal, as is often required on HPVs with GRP bodywork. It is not affected by vibration and is totally waterproof. The Araldite range includes a number of industrial adhesives which are highly recommended for use with GRP. Most high-strength impact adhesives (superglues) can be used on GRP laminates.

Most other adhesives will be incapable of bonding strongly to GRP and should not be used when maximum adhesion is essential.

Core Materials

Core materials, usually polyurethane, are used in sandwich construction, that is basically a laminate consisting of a foam sheet between two or more glass fibre layers. This gives the laminate considerable added rigidity without greatly increasing weight. Foam materials are available which can be bent and folded to follow curved surfaces such as motorcycle parts. Foam sheet can be glued or stapled together, then laminated over to produce a strong box structure, without requiring a mould. Typical formers and core materials are paper rope, polyurethane rigid foam sheet, scoreboard contoured foam sheet, Termanto PVC rigid foam sheet, Term PVC contoured foam sheet and Termino PVC contoured foam sheet.

Formers

A former is anything which provides shape or form to a GRP laminate. They are often used as a basis for stiffening ribs or box sections. A popular material for formers is a paper rope, made of paper wound on flexible wire cord. This is laid on the GRP surface and is laminated over to produce reinforcing ribs, which give added stiffness with little extra weight. The former itself provides none of the extra stiffness; this results entirely from the box section of the laminate rib. Wood, metal or plastic tubing and folded cardboard can all be used successfully as formers. Another popular material is polyurethane foam sheet, which can be cut and shaped to any required form.

Composite Theory

In its most basic form a composite material is one which is composed of two elements working together to produce material properties that are different to the properties of those elements on their own. In practice, most composites consist of a bulk material called the matrix, and a reinforcement material of some kind which increases the strength and stiffness of the matrix.

Polymer matrix composites (PMCs) are the type of composites used in modern vehicle bodywork. This type of composite is also known as fibre-reinforced polymers (or plastics) (FRPs). The matrix is a polymer-based resin and the reinforcement material is a fibrous material such as glass, carbon or aramid. Frequently, a combination of reinforcement materials is used.

The reinforcement materials have high tensile strength, but are easily chaffed and will break if folded. The polymer matrix holds the fibres in place so that they are in their strongest position and protects them from damage.

The properties of the composite are thus determined by:

- The properties of the fibre.
- The properties of the resin.
- The ratio of fibre to resin in the composite – **fibre volume fraction** (FVF).
- The geometry and orientation of the fibres in the composite.

Resin

The choice of resins depends on a number of characteristics, namely:

- Adhesive properties – in relation to the type of fibres being used, and if metal inserts are to be used such as for panel fitting.
- Mechanical properties – particularly tensile strength and stiffness.
- Micro-cracking resistance – stress and age hardening cause the material to crack; the micro-cracks reduce the material strength and eventually lead to failure.
- Fatigue resistance – composites tend to give better fatigue resistance than most metals.
- Degradation from water ingress – all laminates permit very low quantities of water to pass through in a vapour form. If the laminate is wet for a long period, the water solution inside the laminate will draw in more water through the osmosis process.
- Curing properties – the curing process alters the properties of the material. Generally oven curing at between 80°C and 180°C will increase the tensile strength by up to 30%.
- Cost – the different materials have different prices.

The main types of resins are: polyesters, vinylesters, epoxies, phenolics, cyanate esters, silicones, polyurethanes, bismaleides (**BMI**) and polyamides. The first three are those mainly used for manufacturing work as they are reasonably priced. Cyanates, BMI and polyamides cost about 10 times the price of the others.

Reinforcing Fibres

The mechanical properties of the composite material are usually dominated by the contribution of the reinforcing fibres. The four main factors which govern this contribution are:

1   The basic mechanical properties of the fibre.
2   The surface interaction of the fibre and the resin – called the interface.
3   The amount of fibre in the composite – **FVF**.
4   The orientation of the fibres.

The three main reinforcing fibres used in HPVs are glass, carbon and aramid. In addition, the following are used for non-body purposes: polyester, polyethylene, quartz, boron, ceramic and natural fibres, such as jute and sisal.

Aramid fibre is a man-made organic polymer, an aromatic polyamide, produced by spinning fibre from a liquid chemical blend. The bright golden yellow fibres have high strength and low density giving a high specific strength. Aramid has good impact resistance. Aramid is better known by its Dupont trade name Kevlar.

Carbon fibre is produced by the controlled oxidation, carbonisation and graphitisation of carbon-rich organic materials – referred to as precursors – which are in fibre form. The most common precursor is polyacrylonitrile (PAN); pitch and cellulose are also used.

Pre-Impregnated Material (Pre-Preg)

Woven material is available pre-impregnated with resin. It is referred to as pre-preg. This means that the material has exactly the right amount of resin applied to it. The resin fully coats the material – so that there are no dry spots which could lead to component failure. Pre-preg is therefore quicker to use and the resin density is accurate.

Pre-preg has a limited shelf life which is compounded by the fact that it must be stored at $-18°C$. A deep freeze cabinet is therefore needed for storage. The pre-preg cannot be unrolled nor cut when it is in the frozen state, so it must be removed from the freezer and brought up to normal room temperature. It is only possible to freeze and de-frost the pre-preg a limited number of times, so the material must be managed carefully. The usual way to do this is by means of a control card. The dates and times of defrosting are recorded, as is the amount of material taken off the roll. That way the life of the roll and the amount of material left can be seen without removing the roll from the freezer.

Curing

The resin, whether it is by wet lay-up or pre-preg, needs time and heat to dry it out and make it hard. When the hardener is added to the resin it will generate heat chemically. Be careful, this heat can cause fire and other damage. However, at normal temperature, $20°C$, it will take about 5 days for the resin to become fully hard. During this time period the component should not be moved nor should any stress be applied. To speed up the hardening process and to add extra strength to the component it is normal to use an oven. The oven may be a simple box with a heating element, or an autoclave which is a cylindrical shaped oven that can be pressurised or evacuated inside. The normal procedure is to place the newly made component in the oven, or autoclave, then rack up the temperature gently, over a period of about 30 minutes. Maintain the temperature typically at $150°C$ for about 5 hours, then gradually lower the temperature, again over about a 30-minute period. The best way to do this is with a computer control system.

Core Materials

Engineering theory tells us in most cases that the stiffness of a panel is proportional to the cube of its thickness. That is, the further apart that we can keep the outer fibres the stiffer the panel will be. Putting a low-density core between two layers of composite material will add stiffness with minimum weight and at reasonable cost.

Foam

A variety of materials are used, one of the most common is foam. Foam can be made from a variety of synthetic polymers. Densities of foam can vary between 30 and 300 kg/m^3 and thicknesses available are from 5 to 50 mm.

Honeycomb

Honeycombs are made from a variety of materials, including extruded thermoplastic – ABS, polycarbonate, polypropylene and polyethylene – bonded paper, aluminium alloy and, for fire-resistant parts, Nomex. Nomex is a paper-like material based on Kevlar fibres.

Heat

A point to be noted is that most carbon fibre materials are affected by heat. Thermal expansion can lead to micro-cracking. A carbon fibre panel which is painted black will absorb a lot of heat if left in the sun for a long period. This can cause the panel to expand which could lead to micro-cracks in the panel and cracks in the paint work. This will then allow in moisture which will cause further deterioration of the panel.

These provide great strength; but as the components made from them are usually highly labour-intensive, as well as the materials being expensive themselves, the components made from them tend to be very expensive.

Glass-reinforced plastic (GRP) is the cheaper of the composites. GRP is made from very thin glass tubes woven into matting and set in resin. When the GRP dries it has similar properties to aluminium sheeting. It is light and flexible. It can be formed into shapes to make it rigid. Many boats and kit car bodies are made from GRP. Its big advantage is that it can be laid up over a mould to provide the required shape – such as for a car wing or sailing dingy hull. The mould can be used many times.

Carbon fibre is made from strands of carbon material, looks similar to cotton, and is woven into matting and set into resin. It is laid up in the same way as GRP. However, unlike the GRP which can be cured in normal dry air – as in any workshop – carbon fibre needs to be cured in an autoclave – a sort of large oven which is both a vacuum and hot. The vacuum, with the absence of air, is need to prevent air bubbles which could produce weak spots in the component.

Both GRP and carbon fibre components are often made with bonded in metal fixings, usually aluminium alloy for lightness, for the attachment of screws and bolts to connect to other parts. Both materials are also used to sandwich other material to make girder like components.

Tech note

In common terms, GRP is also referred to as glass fibre. Carbon fibre is also referred to simply as carbon.

Work Hardening and Fatigue Failure

We talk about hardening as a good thing; but work hardening is different, it is a bad thing. Work hardening and fatigue failure lead to component breakage. Aluminium and copper both go hard due to time and vibrations. That is, they harden without being noticeably stress loaded. Steel does not do this.

If a steel component is not over-loaded, it will retain its strength for its life time. An example is the use of steel joists in buildings, these will remain straight and true if the building is not over-loaded.

Aluminium will work harden with time and normal vibrations; therefore, it has a finite life span leading to eventual fatigue failure.

Welding and brazing change the structure of the metals at the area of the joint, creating a point more susceptible to failure. The fracture usually occurs about 3 mm from the actual welded area, not at the actual weld itself.

Properties of Materials

Stress – this is usually measured in mega-Pascals (MPa). The load in mega-Newtons (mN) over the cross-sectional area in metres (m). There are several types of stress, metals are usually judged by their ultimate tensile stress (UTS). That is the level of stress at which they will break. Making a component of thicker metal will increase the load which it can carry for any given material.

Hardness and softness – Hardness is the resistance of a material to be indented and bent. Hard materials are usually brittle. The opposite of hardness is softness. For example, high-carbon steel, such as drills and machine tools, are both hard and brittle. Aluminium sheet is both soft and easy to bend.

Ductility – the ability of a material to be drawn, or stretched, without breaking. Copper and silver are examples of ductile metals. When metals are heated they usually become more ductile and can be shaped easily.

Elasticity – materials are elastic if they return to their original shape and size after being stretched. Think of elastic bands. Most materials have an elastic limit, once stretched beyond this point they will not return to their original shape.

Compressive strength – the ability of a material to support loads applied to them. The resistance to being squashed. Components carrying weight, such as mounting blocks, need to resist compressive stress.

Tensile strength – the resistance of a material to being stretched. Bolts and studs are made to resist tensile loads.

Torsional strength – the resistance to being twisted, bolts need high torsional strength as do drive shafts.

Shear strength – the resistance to snapping or tearing. The cutting action of scissors and sheet metal snips is shear, shear strength is resistance to this action, Bolts, clevis pins, gudgeon pins and shackle pins need high shear strength. To give high resistance to shear along with wear resistance it in usual to use a softer material with a hard surface coating.

Brittleness – is when something easily breaks or smashes when a load is applied. Glass and ceramics are very brittle.

Elongation – the amount by which something elongates, grows longer, compared to its original length. Also, the terms **deformation** and **extension** are used where it is not a simple change in length.

Strain – the ration of elongation divided by the original length, usually expressed as a percentage.

Young's modulus – the ration of stress divided by strain.

Strength – usually refers to as the UTS.

Factor of safety – the number of times that the maximum load is compared to the expected load.

Elastic limit – the stress at which a metal does not return to its original shape. Steel is up to a point elastic; you bend it and it bends back. Bend it more and it stays bent.

Stiffness – the load needed to bend an item a particular amount.

Strength to weight – the UTS as a ratio of the density. UTS in MPa, density in gram/cm^3.

Stiffness to weight – the stiffness as a ratio to density.

Aging T numbers – a set of standards that aluminium alloy is hardened to. It is expressed as a working standard giving temperature that the metal is heated to, the length of time it is held at this temperature, and the cooling process.

Welding and brazing dissimilar metals – the welding and brazing of dissimilar metals is possible with modern methods and fluxes. Of course, the joints and parts will have different strengths and properties to those of a normal, single metal, joint. You should check with the material suppliers, and carry out a test joint before using this in a real-life situation.

EN (BSI) standards – European Standards, literally European Norm. British Standards (BS) have merged with these. In America the equivalent is ANSI, Germany has DIN and Japan JIS. There are other equivalents used around the world. The number runs in to hundreds of variations.

Thermal conductivity – the ability of a material to conduct heat, copper is a good conductor of heat, hence its use in the manufacture of cooking pans. Ceramic is resistant to conducting heat.

Electrical conductivity – the ability to transmit electricity, copper and aluminium are used for electrical cables as they are good electrical conductors. PVC is used as an insulator on electrical cable as it resists the flow of electricity, it is an insulator.

Magnetic materials – iron, steel alloys and other compounds containing ferric materials (iron) have magnetic attraction or repulsion. Opposite poles – north and south – attract. Like poles repel.

Fatigue – when a material becomes worn out through usage.

Creep – if a material is placed under load it may stretch or bend over time.

Failure Analysis

Testing

Common types of testing are:

> **Tensile testing** – this involves applying a load to stretch a sample of material, measuring the amount of stretch in relation to the load applied.
>
> **Hardness testing** – this involves forcing either a steel ball, or a steel and diamond cone, into the surface of the sample and measuring the diameter of the indentation in relationship to the force applied. There are several different scales used for this calculation, the two most popular are Brinell and Rockwell.
>
> **Impact testing** – this tests the hardness of a material, its resistance to bend, or in fact fracture. The Izod test is the most popular, this uses a weighted pendulum which is swung at the material sample.
>
> **Creep testing** – a sample material is placed under a constant load at a constant temperature and changes in length, or bending, are recorded against time.
>
> **Fatigue testing** – the sample material, or product, is subjected to a load of the expected type, for instance bending, the stress is plotted on a graph against the number of cycles.
>
> All these tests are carried out until the sample breaks. However, we may wish to test items without damaging them, or in fact to find out if they are damaged, for this we carry out non-destructive testing.

Non-destructive testing (NDT) – forms a major specialism within engineering. A number of companies and laboratories specialise solely in this area of work. By the very nature of the processes involved, and the requirement to be exact and detailed, it is an expensive procedure. There are different types of NDT, these are the most common methods:

- Visual NDT (VT)
- Ultrasonic NDT (UT)
- Radiography NDT (RT)
- Eddy current NDT (ET)
- Magnetic particle NDT (MT)
- Acoustic emission NDT (AE)
- Dye penetrant NDT (PT)
- Leak testing (LT).

During manufacturing the most popular method is visual inspection, a small percentage of the products are taken off the end of the production line and inspected, this usually involves accurate measurement.

For workshop and on-the-job fault finding the use of dye penetrants is the most common as it is the simplest and cheapest. The procedure is as follows:

1. Clean suspected area, removing all oil, grease and other contaminants.
2. Spray – usually from a can – the dye penetrant on suspected area.
3. Leave to dry.
4. Wipe surface with a clean cloth – it will dry with a dusty finish.
5. Any cracks will be visible as the penetrant will still be visible in them.

Radiography and ultrasonic testing have two advantages, firstly the results can be seen on a screen, and secondly they can detect cracks and failures which are hidden under the surface of the item.

Soldering, Brazing and Welding

Engineering construction involves components being joined together. Soldering, brazing and welding are three common methods of joining. In this chapter we discuss these processes.

Comparison of Fusion and Non-Fusion Jointing Processes

The jointing of metals by processes employing fusion of some kind, that is the melting of metal. There are different types of fusion, they may be classified as follows.

Total Fusion

Temperature range: 1130–1550°C approximately. Processes: oxy-acetylene welding, manual metal arc welding, inert gas metal arc welding. In other words, a welded frame, usually without lugs.

Skin Fusion

Temperature range: 620–950°C approximately. Processes: flame brazing, silver soldering, aluminium brazing, bronze welding. A lugged frame or one which is fillet brazed.

Surface Fusion

Temperature range: 183–310°C approximately. Process: soft soldering. Used on lightly loaded components, and for electrical and electronic components.

In total fusion the parent metal and, if used, the filler metal, are both completely melted during the jointing. Tubes can be fused together without additional filler metal being added. Oxy-acetylene welding and manual metal arc welding were the first processes to employ total fusion. In recent years they have been supplemented by methods such as inert-gas arc welding, metal inert-gas (MIG/MAG) and tungsten inert-gas (TIG) welding, carbon dioxide welding and atomic hydrogen welding. Welding is normally carried out at high temperature ranges, the actual temperature being the melting point of the particular metal which is being joined. The parent metal is totally melted throughout its thickness, and in some cases molten filler metal of the correct composition is added by means of rods or consumable electrodes of convenient size. A neat reinforcement weld bead is usually left protruding above the surface of the parent metal, as this gives good mechanical properties in the completed weld. Most metals and alloys can be welded effectively, but there are certain exceptions which, because of their physical properties, are best joined by alternative methods.

In skin fusion, the skin or surface grain structure only of the parent metal is fused to allow the molten filler metal to form an alloy with the parent metal. Hard solders are used in this process, and, as these have greater shear strength than tensile strength, the tensile strength of the joint must be increased by increasing the total surface area between the metals. The simplest method of achieving this is by using a lapped joint in which the molten metal flows between the adjoining surfaces; this accounts for the use of lugs in joining frame tubes. The strength of the joint will be dependent upon the wetted area between the parts to be joined. Skin fusion has several advantages. First, since the filler metals used in these processes have melting points lower than the parent metal to which they are being applied, a lower level of heat is needed than in total fusion, and in consequence distortion is reduced. Second, dissimilar metals can be joined by applying the correct amount of heat to each parent metal, when the skins of both will form an alloy with the molten hard solder. Frame lugs are usually cast from a different alloy of steel to that of the frame tubes. Third, since only the skin of the parent metal is fused, a capillary gap is formed in the lap joint and the molten filler metal is drawn into the space between the surfaces of the metals, this allows easy assembly and wriggle room to adjust the frame angles before making the joint permanent. The filler material, also called spelter, will fill the space available when at the correct temperature to flow.

In surface fusion the depth of penetration of the molten solder into the surfaces to be joined is so shallow that it forms an intermetallic layer which bonds the surfaces together. The process employs soft solders, which are composed mainly of lead and tin. As these also have a low resistance to a tensile pulling force, the joint design must be similar to that of the skin fusion process, i.e. a lapped joint.

Soft and Hard Solders

In spite of the growing use of welding, the techniques of soldering remain among the most familiar in the fabrication of sheet metal articles, and the art of soldering continues to occupy an important place in the workshop. While soldering is comparatively simple, it requires care and skill and can only be learnt by actual experience.

Soldering and brazing are methods of joining components by lapping them together and using a low-melting point alloy so that the parent material is not melted. Soldering as a means of joining metal sheets has the advantage of simplicity in apparatus and manipulation, and with suitable modifications it can be applied to practically all commercial metals.

Soft Soldering

The mechanical strength of soft soldered sheet metal joints is usually in the order of 15–30 MN/m², and depends largely upon the nature of the solder used; the temperature at which the soldering is done; the depth of penetration of the solder, which in turn depends on capillary attraction, i.e. on the power of the heated surface to draw liquid metal through itself; the proper use of correctly designed soldering tools; the use of suitable fluxes; the speed of soldering; and, especially, workmanship.

Solders

Soft solder is an alloy of lead and tin, and is used with the aid of a soldering flux. It is made from two base metals, tin and lead. Tin has a melting point of 232°C and lead 327°C, but the alloy has a lower melting point than either of the two base metals and its lowest melting point is 183°C; this melting point may be raised by varying the percentage of lead or tin in the alloy. A small quantity of antimony is sometimes used in soft solder with a view to increasing its tenacity and improving its appearance by brightening the colour. The small percentage of antimony both improves the chemical properties of the solder and increases its tensile strength, without appreciably affecting its melting point or working properties.

Tech note

There is a great variety of solders, e.g. aluminium, bismuth, cadmium, silver, gold, pewterer's, plumber's, tinman's; solders are usually named according to the purpose for which they are intended.

The following solders are the most popular in use today:

95–100% tin solder is used for high-quality electrical work where maximum electrical conductivity is required, since the conductivity of pure tin is 20–40% higher than that of the most commonly used solders.

60/39.5/0.5 (tin/lead/antimony) solder, the eutectic composition, has the lowest melting point of all tin–lead solders, and is quick setting. It also has the maximum bulk strength of all tin–lead solders, and is used for fine electrical and tinsmith's work.

50/47.5/2.5 (tin/lead/antimony) solder, called tinman's fine, contains more lead and is therefore cheaper than the 60/40 grade. Its properties in terms of low melting range and quick setting are still adequate, and hence it is used for general applications.

45/52.5/2.5 (tin/lead/antimony) solder, known as tinman's soft, is cheaper because of the higher lead content, but has poorer wetting and mechanical properties. This solder is widely used for can soldering, where maximum economy is required, and for any material which has already been tin plated so that the inferior wetting properties of the solder are not critical.

30/68.5/1.5 (tin/lead/antimony) solder, known as plumber's solder, is also extensively used by the car body repairer. Because the material has a wide liquidus–solidus range (about 80°C), it remains in a pasty form for an appreciable time during cooling, and while in this condition it can be shaped or 'wiped' to form a lead pipe joint, or to the shape required for filling dents in frame tubes. Because of its high lead content, its wetting properties are very inferior and the surfaces usually have to be tinned with a solder of higher tin content first.

Fluxes

The function of a flux is to remove oxides and tarnish from the metal to be joined so that the solder will flow, penetrate and bond to the metal surface, forming a good strong soldered joint. The hotter the metal, the more rapidly the oxide film forms. Without the chemical action of the flux on the metal the solder would not tin the surface, and the joint would be weak and unreliable. As well as cleaning the metal, flux also ensures that no further oxidation from the atmosphere which could be harmful to the joint takes place during soldering, as this would restrict the flow of soldering.

Generally, soft soldering fluxes are divided into two main classes: corrosive fluxes and non-corrosive fluxes.

Tech note

Some fizzy drinks contain phosphoric acid – if you drop a dirty coin in the drink the acid will clean it.

Brazing

Brazing is used extensively throughout the frame building trade as a quick and cheap means of joining frame tubes and other components. Although a brazed joint is not as strong as a fusion weld, it has many advantages which make it useful for the frame builder. Brazing is not classed as a fusion process, and therefore cannot be called welding, because the parent metals are not melted to form the joint but rely on a filler material of a different metal of low melting point which is drawn through the joint. The parent metals can be similar or dissimilar as long as the alloy rod has a lower melting point than either of them. The most commonly used alloy is of copper and zinc, which is, of course, brass. Brazing is accomplished by heating the pieces to be joined to a temperature higher than the melting point of the brazing alloy (brass). With the aid of flux, the melted alloy flows between the parts to be joined due to capillary attraction, and actually diffuses into the surface of the metal, so that a strong joint is produced when the alloy cools. Brazing, or hard soldering to give it its proper name, is in fact part fusion and is classed as a skin fusion process.

Brazing is carried out at a much higher temperature than that required for the soft soldering process. A borax type of powder flux is used, which fuses to allow brazing to take place between 750 and 900°C. There are a wide variety of alloys in use as brazing rods; the most popular compositions contain copper in the ranges 46–50% and 58.5–61.5%, the remaining percentage being zinc.

The brazing process comprises the following steps :

1. Thoroughly clean the metal to be joined.
2. Using a welding torch, heat the metals to a temperature below their own critical or melting temperature. In the case of steel, the metal is heated to a dull cherry red.
3. Apply borax flux either to the rod or to the work as the brazing proceeds, to reduce oxidation and to float the oxides to the surface.
4. Use the oxy-acetylene torch with a neutral flame, as this will give good results under normal conditions.

Safety note

When heated, zinc-plated steel (galvanised) gives off very toxic fumes, so full respiratory equipment must be used – it is better to avoid this hazard if possible.

The main advantages of brazing are:

1 The relatively low temperature (750–900°C) necessary for a successful brazing job reduces the risk of distortion.
2 The joint can be made quickly and neatly, requiring very little cleaning up.
3 Brazing makes possible the joining of two dissimilar metals; for example, brass can be joined to steel.
4 It can be used to repair parts that have to be re-chromed. For instance, a chromed fork which has been deeply scratched can be readily filled with brazing and then filed up ready for chroming.
5 Brazing is very useful for joining steels which have a high carbon content, or broken castings where the correct filler rod is not available.

Silver Soldering

Silver solder probably originated in the manufacture and repair of silver-ware and jewellery for the purpose of securing adequate strength and the desired colour of the joint; but the technique of joining sheet metal products and components with silver solder was used for a long time on high-quality bicycle frames. The term 'soft soldering' has been widely adopted when referring to the older process to avoid confusion with the newer hard soldering process, known generally as either silver soldering or silver brazing. The use of silver solder on metals and alloys other than silver has grown largely because of the perfection by manufacturers of these solders which makes them easily applicable to many metals and alloys by means of the oxy- acetylene welding torch.

Solders and Fluxes

Silver solders are more malleable and ductile than brazing rods, and hence joints made with silver solder have a greater resistance to bending stresses, shocks and vibration than those made with ordinary brazing alloys, as you can see this is very appropriate for bicycle frames. Silver solders are made in strip, wire (rod) or granular form and in a number of different grades of fusibility, the melting points varying between 630 and 800°C according to the percentages of silver, copper, zinc and cadmium they contain.

As in all non-fusion processes the important factor is that the joint to be soldered must be perfectly clean. Hence special care must be taken in preparing the metal surfaces to be joined with silver solder. Although fluxes will dissolve films of oxide during the soldering operation, frame tubes and lugs that are clean are much more likely to make a stronger, sounder joint than when impurities are present. The joints should fit closely and the parts must be held together firmly while being silver soldered, because silver solders in the molten state are remarkably fluid and can penetrate into minute spaces between the metals to be joined. The use of a frame building jig is essential with this process. In order to protect the metal surface against oxidation and to increase the flowing properties of the solder, a suitable flux such as borax or boric acid is used.

Silver Soldering Process

In silver soldering the size of the welding tip used and the adjustment of the flame are very important to avoid overheating, as prolonged heating promotes oxide films which weaken both the base metal and the joint material. This should be guarded against by keeping the luminous cone of the flame well back from the point being heated. When the joint has been heated just above the temperature at which the silver solder flows, the flame should be moved away and the solder applied to the joint, usually in rod form. The flame should then be played over the joint so that the solder and flux flow freely through the joint by capillary attraction. The finished silver soldered joint should be smooth, regular in shape and require no dressing up apart from the removal of the flux by washing in water.

When making a silver solder joint between dissimilar metals, concentrate the application of heat on the metal which has the higher heat capacity. This depends on the thickness and the thermal conductivity of the metals. The aim is to heat both members of the joint evenly so that they reach the soldering temperature at the same time.

The most important points during silver soldering are:

1   Cleanness of the joint surfaces
2   Use of the correct flux
3   The avoidance of overheating.

Aluminium Brazing

There is a distinction between the brazing of aluminium and the brazing of other metals. For aluminium, the brazing alloy is one of the aluminium alloys having a melting point below that of the parent metal. For other metals, the brazing alloys are often based on copper–zinc alloys (brasses – hence the term brazing) and are necessarily dissimilar in composition to the parent metal.

Wetting and Fluxing

When a surface is wetted by a liquid, a continuous film of the liquid remains on the surface after draining. This condition, essential for brazing, arises when there is mutual attraction between the liquid flux and solid metal due to a form of chemical affinity. Having accomplished its primary duty of removing the oxide film, the cleansing action of the flux restores the free affinities at the surface of the joint faces, promoting wetting by reducing the contact angle developed between the molten brazing alloy and parent metal. This action assists spreading and the feeding of brazing alloy to the capillary spaces, leading to the production of well-filled joints. An important feature of the brazing process is that the brazing alloy is drawn into the joint area by capillary attraction: the smaller the gap is between the two metal faces to be joined, the deeper is the capillary penetration.

The various grades of pure aluminium and certain alloys are amenable to brazing. Aluminium–magnesium alloys containing more than 2% magnesium are difficult to braze, as the oxide film is tenacious and hard to remove with ordinary brazing fluxes. Other alloys cannot be brazed because they start to melt at temperatures below that of any available brazing alloy. Aluminium–silicon alloys of nominal 5%, 7.5% or 10% silicon content are used for brazing aluminium and the alloy of aluminium and 1.5% manganese.

The properties required for an effective flux for brazing aluminium and its alloys are as follows:

1   The flux must remove the oxide coating present on the surfaces to be joined. It is always important that the flux be suitable for the parent metal, but especially so in the joining of aluminium–magnesium alloys.

2　It must thoroughly wet the surfaces to be joined so that the filler metal may spread evenly and continuously.

3　It must flow freely at a temperature just below the melting point of the filler metal.

4　Its density, when molten, must be lower than that of the brazing alloy.

5　It must not attack the parent surfaces dangerously in the time between its application and removal.

6　It must be easy to remove from the brazed assembly.

Many types of proprietary fluxes are available for brazing aluminium. These are generally of the alkali halide type, which are basically mixtures of the alkali metal chlorides and fluorides. Fluxes and their residues are highly corrosive and therefore must be completely removed after brazing by washing with hot water.

Brazing Method

When the cleaned parts have been assembled, brazing flux is applied evenly over the joint surface of both parts to be brazed and the filler rod (brazing alloy). The flame is then played uniformly over the joint until the flux has dried and become first powdery, then molten and transparent. (At the powdery stage care is needed to avoid dislodging the flux, and it is often preferable to apply flux with the filler rod.) When the flux is molten the brazing alloy is applied, preferably from above, so that gravity assists in the flow of metal. In good practice the brazing alloy is melted by the heat of the assembly rather than directly by the torch flame. Periodically, the filler rod is lifted and the flame is used to sweep the liquid metal along the joint; but if the metal is run too quickly in this way it may begin to solidify before it properly diffuses into the mating surfaces. Trial will show whether more than one feed point for the brazing alloy is necessary, but with proper fluxing, giving an unbroken path of flux over the whole joint width, a single feed is usually sufficient.

Bronze Welding

Bronze welding is carried out much as in fusion welding except that the base metal is not melted. The base metal is simply brought up to tinning temperature (dull red colour) and a bead is deposited over the seam with a bronze filler rod. Although the base metal is never actually melted, the unique characteristics of the bond formed by the bronze rod are such that the results are often comparable with those secured through fusion welding. Bronze welding resembles brazing, but only up to a point. The application of brazing is generally limited to joints where a close fit or mechanical fastening serves to consolidate the assembly and the joint is merely strengthened or protected by the brazing material. In bronze welding the filler metal alone provides the joint strength, and it is applied by the manipulation of a heating flame in the same manner as in gas fusion welding. The heating flame is made to serve the dual purpose of melting off the bronze rod and simultaneously heating the surface to be joined. The operator in this manner controls the work: hence the term 'bronze welding'.

Almost any copper–zinc alloy, copper–tin alloy or copper–phosphorus alloy can be used as a medium for such welding, but the consideration of costs, flowing qualities, strength and ductility of the deposit have led to the adoption of one general purpose 60–40 copper–zinc alloy with minor constituents incorporated to prevent zinc oxide forming and to improve fluidity and strength. Silicon is the most important of these minor constituents, and its usefulness is apparent in three directions. First, in the manner with which it readily unites with oxygen to form silica, silicon provides a covering for the molten metal which prevents zinc volatilisation and serves to maintain the balance of the constituents of the alloy; this permits the original high strength of the alloy to be carried through to the deposit. Second, this coating of silica combines with the flux used in bronze welding to form a very fusible slag, and this materially assists the tinning operation, which is an essential feature of any bronze welding process. Third, by its capacity for retaining gases in solution during solidification of the alloy, silicon prevents the formation of gas holes and porosity in the deposited metal, which would naturally reflect unfavourably upon its strength as a weld.

It is essential to use an efficient and correct flux. The objects of a flux are: first, to remove oxide from the edges of the metal, giving a chemically clean surface on to which the bronze will flow, and to protect the heated edges from the oxygen in the atmosphere; second, to float oxide and impurities introduced into the molten pool to the surface, where they can do no harm. Although general-purpose fluxes are available, it is always desirable to use the fluxes recommended by the manufacturer of the particular rod being employed.

Bronze Welding Procedure

1   An essential factor for bronze welding is a clean metal surface. If the bronze is to provide a strong bond, it must flow smoothly and evenly over the entire weld area. Clean the surfaces thoroughly with a stiff wire brush. Remove all scale, dirt or grease, otherwise the bronze will not adhere. If a surface has oil or grease on it, remove these substances by heating the area to a bright red colour and thus burning them off.

2   On thick sections, especially in repairing castings, bevel the edges to form a 90° V-groove. This can be done by chipping, machining, filing or grinding.

3   Adjust the torch to obtain a slightly oxidising flame. Then heat the surfaces of the weld area.

4   Heat the bronzing rod and dip it in the flux. (This step is not necessary if the rods have been pre-fluxed.) In heating the rod, do not apply the inner cone of the flame directly to the rod.

5   Concentrate the flame on the starting end until the metal begins to turn red. Melt a little bronze rod on to the surface and allow it to spread along the entire seam. The flow of this thin film of bronze is known as the tinning operation. Unless the surfaces are tinned properly the bronzing procedure to follow cannot be carried out successfully. If the base metal is too hot, the bronze will tend to bubble or run around like drops of water on a warm stove. If the bronze forms into balls which tend to roll off, just as water would if placed on a greasy surface, then the base metal is not hot enough. When the metal is at the proper temperature the bronze spreads out evenly over the metal.

6   Once the base metal is tinned sufficiently, start depositing the proper size beads over the seam. Use a slightly circular torch motion and run the beads as in regular fusion welding with a filler rod. Keep dipping the rod in the flux as the weld progresses forward. Be sure that the base metal is never permitted to get too hot.

7   If the pieces to be welded are grooved, use several passes to fill the V. On the first pass make certain that the tinning action takes place along the entire bottom surface of the V and about half-way up on each side. The number of passes to be made will depend on the depth of the V. When depositing several layers of beads, be sure that each layer is fused into the previous one.

Health and Safety and the Environment

The materials used in soldering can be very dangerous if not used in a safe way, as well as the normal workshop precautions the following should be especially noted

- All soldering and brazing must be carried out in an area with suitable fume extraction.
- Fluxes are mainly corrosive and should be handled accordingly.
- Always wash your hands after soldering or brazing.
- Use PPE as directed by your company safety guide lines.
- All materials must be stored securely and not be accessible to unauthorised personnel.

> **Safety note**
>
> You must remember that all gases are dangerous if not handled and stored correctly – always follow the manufacturer's instructions.

The following is a summary of gas characteristics and cylinder colour codes.

Oxygen

Cylinder colour: black.

Characteristics: no smell. Generally considered non-toxic at atmospheric pressure. Will not burn but supports and accelerates combustion. Materials not normally considered combustible may be ignited by sparks in oxygen-rich atmospheres.

Nitrogen

Cylinder colour: grey with black shoulder. Characteristics: no smell. Does not burn. Inert, so will cause asphyxiation in high concentrations.

Argon

Cylinder colour: blue.

Characteristics: no smell. Heavier than air. Does not burn. Inert. Will cause asphyxiation in absence of sufficient oxygen to support life. Will readily collect in the bottom of a confined area. At high concentrations, almost instant unconsciousness may occur followed by death. The prime danger is that there will be no warning signs before unconsciousness occurs.

Propane

Cylinder colour: bright red and bearing the words: 'Propane' and 'Highly flammable'.

Characteristics: distinctive fish-like offensive smell. Will ignite and burn instantly from a spark or piece of hot metal. It is heavier than air and will collect in ducts, drains or confined areas. Fire and explosion hazard.

Acetylene

Cylinder colour: maroon.

Characteristics: distinctive garlic smell. Fire and explosion hazard. Will ignite and burn instantly from a spark or piece of hot metal. It is lighter than air and less likely than propane to collect in confined areas. Requires minimum energy to ignite in air or oxygen. Never use copper or alloys containing more than 70% copper or 43% silver with acetylene.

Hydrogen

Cylinder colour: bright red.

Characteristics: no smell. Non-toxic. Much lighter than air. Will collect at the highest point in any enclosed space unless ventilated there. Fire and explosion hazard. Very low ignition energy.

Carbon Dioxide

Cylinder colour: black, or black with two vertical white lines for liquid withdrawal.

Characteristics: no smell but can cause the nose to sting. Harmful. Will cause asphyxiation. Much heavier than air. Will collect in confined areas.

Argoshield

Cylinder colour: blue with green central band and green shoulder.

Characteristics: no smell. Heavier than air. Does not burn. Will cause asphyxiation in absence of sufficient oxygen to support life. Will readily collect at the bottom of confined areas.

Safety Measures

General gas storage procedures

1   Any person in charge of storage of compressed gas cylinders should know the regulations covering highly flammable liquids and compressed gas cylinders as well as the characteristics and hazards associated with individual gases.
2   It is best to store full or empty compressed gas cylinders in the open, in a securely fenced compound, but with some weather protection.
3   Within the storage area oxygen should be stored at least 3 m from fuel gas supply.
4   Full cylinders should be stored separately from the empties, and cylinders of different gases, whether full or empty, should be segregated from each other.
5   Other products must not be stored in a gas store, particularly oils or corrosive liquids.
6   It is best to store all cylinders upright, taking steps, particularly with round-bottomed cylinders, to see that they are secured to prevent them

from falling. Acetylene and propane must *never* be stacked horizontally in storage or in use.

7   Storage arrangements should ensure adequate rotation of stock.

Acetylene Cylinders

1   The gas is stored together with a solvent (acetone) in maroon-painted cylinders, at a pressure of 17.7 bar maximum at 15°C. The cylinder valve outlet is screwed left-handed.

2   The hourly rate of withdrawal from the cylinder must not exceed 20% of its content.

3   Pressure gauges should be calibrated up to 40.0 bar.

4   As the gas is highly flammable, all joints must be checked for leaks using soapy water.

5   Acetylene cylinders must be stored and used in an upright position and protected from excessive heat and coldness.

6   Acetylene can form explosive compounds in contact with certain metals and alloys, especially those of copper and silver. Joint fittings made of copper should not be used under any circumstances.

7   The colour of cylinders, valve threads, or markings must not be altered or tampered with in any way.

Oxygen Cylinders

1   This gas is stored in black painted cylinders at a pressure of 200/230 bar maximum at 15°C.

2   Never under any circumstances allow oil or greases to come into contact with oxygen fittings because spontaneous ignition may take place.

3   Oxygen must not be used in place of compressed air.

4   Oxygen escaping from a leaking hose will form an explosive mixture with oil or grease.

5   Do not allow cylinders to come into contact with electricity.

6   Do not use cylinders as rollers or supports.

7   Cylinders must not be handled roughly, knocked or allowed to fall to the ground.

General Equipment Safety

All equipment should be subjected to regular periodic examination and overhaul. Failure to do so may allow equipment to be used in a faulty state, and may be dangerous.

Rubber hose Use only hose in good condition, fitted with the special hose connections attached by permanent ferrules. Do not expose the hose to heat, traffic, slag, sparks from welding operations, or oil or grease. Renew the hose as soon as it shows any sign of damage.

Pressure regulators Always treat a regulator carefully. Do not expose it to knocks, jars or sudden pressure caused by rapid opening of the cylinder valve. When shutting down, release the pressure on the control spring after the pressure in the hoses has been released. Never use a regulator on any gas except that for which it was designed, or for higher working pressures. Do not use regulators with broken gauges.

Welding torch When lighting up and extinguishing the welding torch, the manufacturer's instructions should always be followed. To clean the nozzle use special nozzle cleaners, never a steel wire.

Fluxes Always use welding fluxes in a well-ventilated area.

Goggles These should be worn at all times during welding, cutting or merely observing.

Protection Leather or fire-resistant clothing should be worn for all heavy welding or cutting. The feet should be protected from sparks, slag or cut material falling on them.

Gas Shielded Arc Welding (MIG, MAG and TIG)

Development of Gas Shielded Arc Welding

Originally this process was evolved in America in 1940 for welding in the aircraft industry. It developed into the tungsten inert-gas shielded arc process, which in turn led to shielded inert-gas metal arc welding. The latter became established in the United kingdom in 1952.

In the gas shielded arc process, heat is produced by the fusion of an electric arc maintained between the end of a metal electrode, either consumable or non-consumable, and the part to be welded, with a shield of protective gas surrounding the arc and the weld region. There are at present in use three different types of gas shielded arc welding:

Tungsten inert gas (TIG) The arc is struck by a non-consumable tungsten electrode and the metal to be welded, and filler metal is added by feeding a rod by hand into the molten pool.

Metal inert gas (MIG) This process employs a continuous feed electrode which is melted in the intense arc heat and deposited as weld metal: hence the term consumable electrode. This process uses only inert gases, such as argon and helium, to create the shielding around the arc.

Metal active gas (MAG) This is the same as MIG except that the gases have an active effect upon the arc and are not simply an inert envelope. The gases used are carbon dioxide or argon/carbon-dioxide mixtures.

Tech note

Gas tungsten arc welding (GTAW) This is the terminology used in America and many parts of Europe for the TIG welding process, and it is becoming increasingly accepted as the standard terminology for this process.

Cutting Speeds

Cutting speeds of materials vary with both the material and the tools used (Figure 8.1 and Table 8.1).

Cutting Speed = Surface Speed

D mm diameter of work
N rpm spindle speed of lathe, etc
S m/m cutting speed

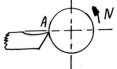

Distance moved by point A over the surface in one rev is:

$$\pi D \text{ mm} = \frac{\pi D}{1000} \text{ m}$$

Cutting speed $S = \dfrac{\pi D N}{1000}$

$$N = \frac{1000 S}{\pi D}$$

$$D = \frac{1000 S}{\pi N}$$

When drilling or milling D is tool diameter.

Figure 8.1 Formula for cutting speeds.

Table 8.2 Cutting Speeds in Metres per Minute (MPM) using High-Speed Steel
Tools – HSS

Material	Turning	Drilling	Milling
Mild steel	20–35	20–30	20–35
Stainless steel	20–35	15–25	60–115
Aluminium	150–180	30–120	150–180
Brass	90–105	50–55	90–105
Brass and bronze	60–75	35–40	60–75
Copper and phosphorus bronze alloys	30–35	20	30–35
Wood	60–150	20–90	120
Plastics – machinable	60–150	20–90	120

Always check the speeds, and other information, suggested by the material and the tool suppliers. When first using a new material, or tool, it is advisable to take a trial cut on a sample of material, before starting production.

Skills and Questions

1. List the three main classifications of carbon steel.
3. What is the difference between curing GRP and carbon fibre?
4. Name the ore used for making aluminium, and state how it is converted into aluminium.
5. What is vegan leather?
6. State the property added to steel by chromium.
7. Differentiate between the terms compression and tensile stress.
8. Describe one method of NDT.
9. What is an Izod test used for, and how is it applied?
10. Sketch any mechanical or electrical item and indicate the materials used for each of the parts.

Chapter 9

Mechanical Principles

See also Chapter 7 on engineering science.

Vectors and Forces

In November 2021, NASA launched a rocket referred to as the Double Asteroid Redirection Test (DART), it weighed 570 kg. It was aimed at the Dimorphos asteroid that was thought may crash into Earth. The rocket that was travelling at 14,000 mph hit the asteroid in September 2022. It was 6.8 million miles from Earth. The vector force of DART knocked the asteroid to change its course. The vector of the asteroid was changed.

Co-planar forces – that is forces in a single flat plane – a sheet of paper is a flat plane, not three-dimensional (3D), helps us to understand how changing one vector quantity will affect others. This experiment will help you to understand it.

Experiment: Investigating the Condition for Equilibrium of Three Co-Planar Forces Acting Through a Point

Objective – to show that when three forces acting through a point on the same plane are in equilibrium, they can be represented by a closed triangle called the triangle of forces.

Apparatus – see Figure 9.1 – this comprises of:

- Two retort stands and clamps
- Two G-clamps, two force meters (spring balances) of the range 0–10 Newton
- Large ruler
- Large protractor
- Set of 100 gram (10 Newton) slotted masses (weights) with hanger
- Length of string

DOI: 10.1201/9781003284833-9

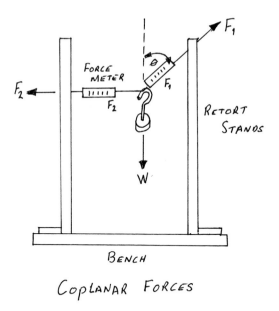

Figure 9.1 Coplanar forces apparatus.

Procedure

1. The apparatus was set up as in Figure 9.1.
2. A 500 g mass, made up of 4 × 100 g weights and hanger, giving in effect 5 N, was added for the weight W.
3. The set-up was adjusted so that the force meter F_2 was horizontal.
4. The readings of force meters F_1 and F_2 were recorded.
5. The angle Θ (theta) on the protractor was noted.
6. The procedure was repeated for three more weights W of 600 g, 700 g and 800 g.

Results

Mass (g)	Weight (N)	F_2 Horizontal (N)	F_1 Inclined (N)	Θ Angle (degrees)
500	5	1.9	5.5	20
600	6	2.5	6.6	25
700	7	4.2	8.3	30
800	8	4.9	9.4	35

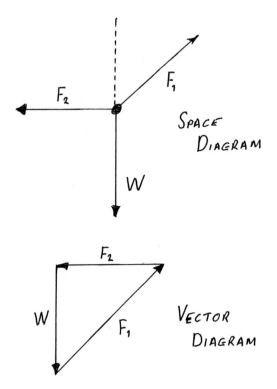

Figure 9.2 Vector diagram.

Analysis and Conclusion

See Figure 9.2.

These results may be analysed by scale drawings or mathematically. The formulas for the mathematical analysis are:

- $F_2 = F_1 \sin \Theta$
- $W = F_1 \cos \Theta$

Table of analysed results

F_2	$F_1 \sin \Theta$	Difference	W	$F_1 \cos \Theta$	Difference
1.9	1.83	0.07	5	5.16	0.16
2.5	2.73	0.23	6	5.93	0.07
4.2	4.15	0.05	7	7.13	0.13
4.9	4.75	0.15	8	7.7	0.3

The results are reasonably accurate within experimental error, possible reasons for inaccuracy are:

- Inaccuracy of force meters, spring operating equipment tends to be more accurate for mid-range readings and less accurate at each end of the scale.
- Inaccuracy of taking the reading – parallax error with the wide markings contribute to error.
- Using g with a value of 10 as against 9.81 will give a constant numerical error.
- The possibility that the horizontal component was not truly horizontal.

Using trigonometry and Pythagoras – to resolve forces in equilibrium follow the normal rules for resolving triangles – see Chapter 6 on calculations. Given a right angle triangle, with the following forces – F:

- Hypotenuse FR
- Opposite FV
- Adjacent FH
- Angle Θ (theta)

$$\text{Sin}\,\Theta = \frac{FV}{FR}$$

$$\text{Cos}\,\Theta = \frac{FH}{FR}$$

$$\text{Tan}\,\Theta = \frac{FV}{FH}$$

Or using Pythagoras:

$$FR^2 = FV^2 + FH^2$$

When the forces are in equilibrium the sum of the vertical forces will equal zero and the sum of the horizontal forces will equal zero.

$$\Sigma F_{vertical} = 0$$

$$\Sigma F_{vertical} = 0$$

Tech note

Σ (capital sigma) indicates sum of.

Simply Supported Beams

Beams are widely used in engineering and construction. Sometime we can clearly see them as girders, or rolled steel joists (RSJs) in buildings: other times they are not recognisable, nor clearly visible, such as floor structures in cars. For a beam to be stable in equilibrium, the clockwise – called positive – moments must equal the negative anti-clockwise moments.

Tech note

A moment is a torque, or turning force, multiplied by length about a point. It is typically measured in newtons times by metres – Nm.

Referring to the diagrams.

(a)
Take moments about R_1.
As the beam is in equilibrium the sum must be equal to zero (0).
The sum of the moments about R_1

$$0 = (100 \times 3) - (R_2 \times 4) + (40 \times 6)$$

Therefore $R_2 = \dfrac{300 + 240}{4}$

$$R_2 = 135$$

(b)
A uniformly distributed load, indicated by a wavy top line in a diagram, is calculated by finding the total load and assuming that it is in the middle of the area. 200 N/m over a length of 4 m is calculated to give a total load of:

$$200 \times 4 = 800 \text{ N}$$

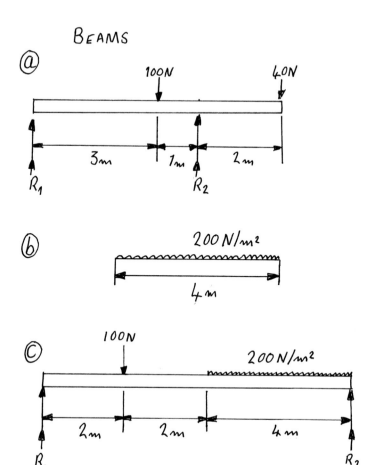

Figure 9.3 Beams.

(c)
Taking moments about R_1

$$0 = (100 \times 2) + (200 \times 4 \times 6) - (R_2 \times 8)$$

$$R_2 = \frac{(100 \times 2) + (200 \times 4 \times 6)}{8}$$

$$R_2 = 625 \text{ N}$$

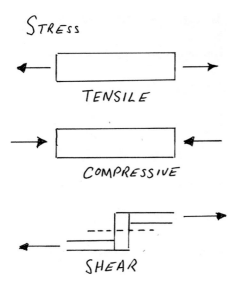

Figure 9.4 Stress and strain.

Stress and Strain

Stress is the load applied directly on a surface expressed as load per unit area. It is said to be the load normal to the surface – this means at 90° to the surface. The force is given in pascals (Pa), which is the equivalent of 1 Newtons (N) and the area of 1 square metre (m^2)

$1\ N/m^2 = 1\ Pa$

Direct stress (sigma σ) = normal force/cross-sectional area

Shear stress (tau τ) = shear force/shear area

Tech note

Normal in mathematics and science means at a right angle (90°) to another surface – plainly it is upright.

Strain is the amount by which the item is stretched when in tension, or compressed when in compression, expressed as a percentage of the original size.

strain = change in length/original length

This is often expressed as a percentage, %

Tech note

See Young's modulus in Chapter 7.

Jean le Rond d' Alembert, an 18th-century French polymath, was interested to find out why ropes broke when lifting bags of grain, and other produce, into the storage lofts of shops and workshops from the streets of France. The rope looked strong enough for the job; but often broke for no apparent reason. He discovered that the rope could carry the weight, and would support the load fully. However, he discovered that if the rope was snatched, in other words caused to try to accelerate the load quickly, it would snap. He discovered:

Under static conditions the load on the rope, in other words force (f_1) was equal to mass × gravity.

$$F_1 = mg$$

When the load was moved quickly from rest, snatched, another accelerating force (F_2) was applied, this was equal to mass × acceleration.

$$F_2 = ma$$

So the total force on the rope is $F_1 + F_2$

Total force = $F_1 + F_2$

If the load (M) is 1000 N, gravity is 9.81 ms²

$$F_1 = 1000 \times 9.81 = 9810 \text{ N}$$

If the rope is pulled hard to give an acceleration of 4 ms²

$$F_2 = 1000 \times 4 = 4000 \text{ N}$$

Total $F_1 + F_2$ = 9810 N + 4000 N

$$= 13,810 \text{ N}$$

JEAN le ROND d' ALEMBERT

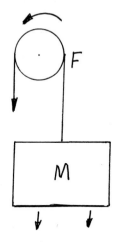

$$F_1 = mg$$

$$F_2 = mga$$

$$\text{TOTAL} = F_1 + F_2$$

Figure 9.5 Lifting load with pulley.

Tech note

If you are towing a car with a rope, providing that the speed of both vehicles is kept constant the rope will stay intact. However, if the towing vehicle suddenly accelerates, the rope may easily snap. This is because the accelerating force has now been added to the constant force.

Momentum – a body, such as a car on a road, possesses mass in kg and velocity in m/s. It will carry on travelling with this mass and velocity until another force prevents it from doing so. In the case of a car on a flat straight road with the engine turned off (which it should be noted is illegal) the air resistance will eventually stop it.

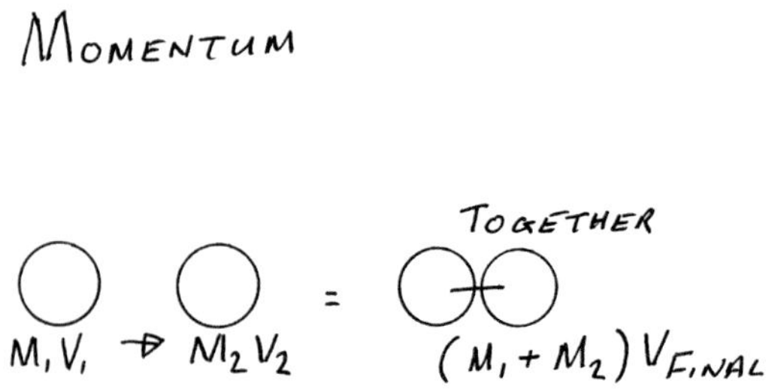

MOMENTUM

Figure 9.6 Momentum example.

If there are two railway wagons, each will have its own momentum, and if one runs into the other, when on a track with negligible resistance, the two will now possess a new momentum.

Momentum = mass × velocity

Momentum = mass (m) × velocity (v)

$$= kg \times m/s$$

For example, wagon 1, mass 300 kg, is travelling at 14 m/s and runs into wagon 2, of 200 kg, travelling in the same direction at a velocity of 12 m/s. Calculate the new velocity.

$(m_1 \, v_1) + (m_2 \, v_2) = (m_1 + m_2) \, v_{final}$

$(300 \times 14) + (200 \times 10) = (300 + 200) \, v_{final}$

$6200 = 500 \, v_{final}$

$v_{final} = 12.4$ m/s

Skills and Questions

1. Carry out the experiment investigating the condition for equilibrium of three co-planer forces acting through a point.

Chapter 10

Electrical and Electronic Principles

WARNING – electricity is dangerous. You cannot see it but it can kill you. There is no such thing as a safe voltage, people have been killed by electrocution from a small torch battery.

See also Chapter 7 on engineering science.
Electricity is the flow of electrons in a circuit. An electron is a very small part of an atom; an atom is the smallest particle of any material. Some materials, such as copper and aluminium, have atoms which have loosely attached electrons. With these materials the electrons can easily flow – we call these conductors. Other materials, such as rubber and ceramic, have their electrons tightly attached to their atoms, we call these non-conductors – or insulators.

Electricity, in other words the movement of electrons, flows between materials that are in a different state of charge. This means from those that have a higher state of charge to one with a lesser charge. Let's have a look the definitions of some of the technical terms used in electric work.

Volt – the unit of potential difference, also referred to as electrical pressure or electromotive force (EMF). The volt is practically defined in Ohm's law as the EMF necessary to produce a current of 1 amp through a resistance of 1 ohm.

Amp – full name ampere, is the unit of current and is defined as the flow of one coulomb of electricity per second. You could think of it as volume.

Coulomb – defined as a base unit of electricity, it is calculated as 6.241×10^{18} electrons. It is, in effect, the mass of electricity.

Ohm – the unit of resistance in a circuit in which a potential difference of one volt produces a current of one amp. Some materials offer more resistance to the flow of electricity than others.

The effects of an electric current – the flow of electricity brings about a number of effects on anything that it flows through. These are:

DOI: 10.1201/9781003284833-10

- **Heating effect** – anything that electricity flows though will get warmer. If you try to pass too much electricity through a wire, or other component, it will get too hot and burn. Fuses work on the basis that if too much current is passed through them they will melt and break the electrical circuit. Hair dryers work on the electricity getting a heating element warm to blow out warm air.
- **Magnetic effect** – the movement of electrons causes the generation of a magnetic field. Electromagnets are used in switching and for sorting ferrous and non-ferrous materials in recycling centres.
- **Chemical effect** – passing electricity through a chemical can change the formulation of the chemical. A car battery is charged up by passing electricity through it, this changes the chemical composition. Processes such as electroplating require electricity to move the metal from the liquid to the item to be coated.

Relative resistivity – different metals are used in the construction of conductors; they have different capabilities of conductivity.

However, there are other reasons for choosing different metals. Aluminium is light in weight, so a good choice when this is an important factor – such as for overhead power lines. Silver and gold are noble metals, they do not affect the human body; but they are in short supply – so very expensive. Copper is very ductile, so that it can be drawn into thin wire easily. Brass is easy to cast and machine, so it is used for connectors and terminations.

Temperature coefficients – the resistivity of materials changes with temperature. For most materials, as the temperature increases the resistance increases, however there are some materials of which the resistance decreases with an increase in temperature, these are referred to as negative temperature co-efficient (NTC). NTC materials are often used in devices that operate switches and gauges. You see the needle rising on a temperature gauge as the NTC device allows more electricity to flow.

Temperature coefficients are usually given as a figure at zero degrees Celsius, or 273 kelvin (k). You then simply multiply the coefficient by the

Table 10.1 Relative Resistivity of Different Metals

	Metal	Resistivity at 20°C – 10–8 Ohms per metre
1	Silver	1.59
2	Copper	1.72
3	Gold	2.44
4	Aluminium	2.82
5	Nickel	6.99
6	Brass	8

Table 10.2 Temperature Coefficients for Resistivity at 0°C

	Metal	Temperature coefficients at 0°C per degree C
1	Silicone	−0.075
2	Germanium	−0.048
3	Carbon	−0.0005
4	Gold	0.0034
5	Silver	0.0038
6	Copper	0.0039
7	Aluminium	0.0040
8	Nickel	0.0062

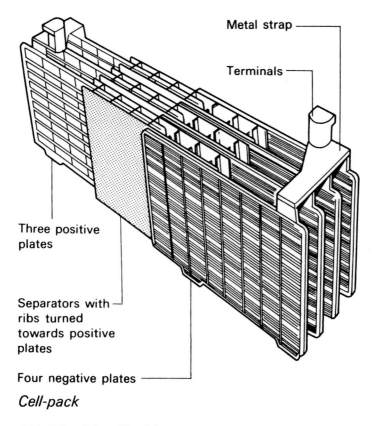

Metal strap

Terminals

Three positive plates

Separators with ribs turned towards positive plates

Four negative plates

Cell-pack

Figure 10.1 Cell pack for a 12-volt battery.

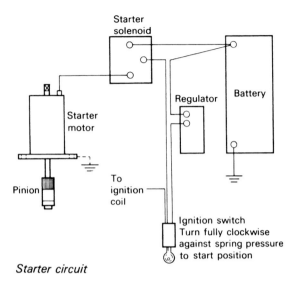

Starter circuit

Figure 10.2 Wiring diagram for a vehicle starter motor.

number of degrees above this base temperature. For NTC materials the result will be a minus figure.

Volt drop – potential difference. Batteries and other electrical power sources are often referred to by their nominal voltage. That is a voltage rating that is accepted for that item, for instance a 12-volt car, or motor-cycle battery. In actual fact, when a typical 12-volt battery is fully charged it will have six cells each giving out 2.2 volts, so a total of 13.2 volts. However, when any electrical circuit is switched on – for instance the car's lights – the voltage reading across the battery terminals will go down. This is because of the internal resistance of the battery giving a resistance to the circuit. When you are starting a petrol/diesel-engined car the volt drop will be much greater because of the amperage taken from the battery by the starter motor.

Tech note

A simple way to test a battery is to measure the volt drop on starting. Clip a multi-meter across the positive and negative terminals, watch the voltage as you start the engine. If the voltage drops below 10.5 V then the battery is likely to be about to completely fail.

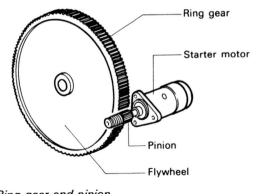

Ring gear and pinion

Figure 10.3 How a starter motor drives the engine.

Direct current (DC) and **alternating current (AC)** – the electrical circuits used in most electrical/electronic items, such as radios, cars and phones, are DC, that is they have a positive (+) and negative (–) circuit; the voltage will usually be less than 12 volt. The power supply to your home however is most likely AC, that is it alternates between the terminals, it is usually between 230 and 240 volt.

Three-phase – when a lot of power is needed, such as in a factory, the supply will be three-phase of about 415 volt. Three-phase uses three cables with phase angles 120° apart.

Grid supply – the power supply from the national grid is usually at about 11 kV (11,000 V) this is three-phase; you can usually see sets of three wires on pylons. The reason for the high voltage is that it is more efficient to transmit power at a high voltage when long distances are to be covered. Remember power = volts × amps, so more volts means less amps for the same amount of power. Less amps means thinner cable can be used.

> **Revision** – in a nut shell, the electron is responsible for the flow of electricity, referred to as a current, and for all practical purposes the current is assumed to flow from the positive terminal of a battery or power source to the negative terminals when the electrons are acted on by a force. The current produced in this way is measured in amperes and the force acting on the electrons in measured in volts. These terms, amperes and volts, together with the ohm, which is the unit of resistance to a current, are defined separately.

Circuit – any unbroken path of conductors along which an electric current flows is called a circuit.

Open circuit – when there is an unintentional break in a circuit it is said to be open. For example, a broken wire, or disconnected terminal, preventing the flow of electricity, even when it is switched on, is called an open circuit.

Short circuit, or **dead short**, is when the current carrying supply is shorted to the device chassis or negative side wiring. This is usually caused by the insulation of the supply wire being chaffed. Dampness can cause short circuits, or partial shorts. Water is a conductor, batteries stored in damp conditions can become flat by discharging slowly through the damp surroundings.

Ohm's law – Georg Simon Ohm was a 19th-century German school teacher who discovered the relationship between volts and amps, known as Ohm's law. The quantity of current, in amperes, flowing in a circuit is proportional to the pressure, in volts, divided by the resistance in ohms.

I = current in amperes

V = pressure in volts (also referred to as E for EMF)

R = resistance in ohms

$I = V/R$

$R = V/I$

$V = I \times R$

Series and Parallel Circuits

Series circuits are where the resistances are connected end to end, in daisy chain form; the characteristics of a series circuit are:

- The value of the current flowing through each resistance is the same.
- The total voltage is divided between the resistances in proportion to their values. That is, the higher the resistance in ohms the greater the voltage across this resistance.
- Any resistance placed in a series circuit will increase the resistance of the circuit and reduce the current.
- Should one resistor fail, as in a string of decorative lights, the whole circuit will fail.

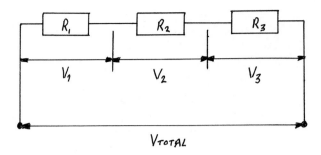

Figure 10.4 Resistors in series.

The total voltage flowing in a series circuit is equal to the sum of the voltages across each of the resistors:

$$V_{total} = V_1 + V_2 + V_3 \ldots\ldots\text{etc. to the total number}$$

The same applies to the total resistance of the circuit:

$$R_{total} = R_1 + R_2 + R_3 \ldots\ldots\ldots\ldots \text{etc. to the total number}$$

Parallel circuits are where the resistances are connected positive to positive and negative to negative, in raster, or parade ground fashion; the characteristics of a parallel series circuit are:

- The value of the voltage across each of the resistances is equal.
- The total current flowing is divided between the resistances and is inversely proportional to their value. That is, the higher the value of the resistance, the less the current in amperes flowing through that resistance.
- Adding resistances in parallel reduces the total resistance of the circuit.

When resistors are connected in parallel, the voltage V across each one is the same.
The sum of the currents flowing through each resistor will be equal to the total circuit current I.

$$I_{total} = I_1 + I_2 + I_3 \ldots\ldots\ldots \text{etc. to the total number}$$

Because the voltage is the same across each of the resistors, from Ohm's law:

$$I = V/R$$

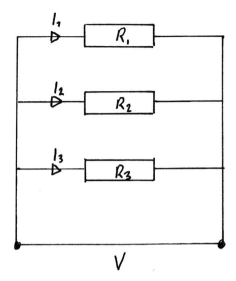

Figure 10.5 Resistors in parallel.

We can deduce that:

$I_1 = V/R_1 + V/R_2 + V/R_3$ ..... etc. to the total number

Dividing by V gives:

$1/R_{total} + 1/R_1 + 1/R_2 + 1/R_3$ .... etc. to the total number

Series and parallel circuit combinations – calculating analysis for these circuits entails breaking them down into small parts and carrying out calculations for each section.

Tech note

When studying circuits in detail it is normal to use a simulator software package, or app. This allows you to construct the circuit and take appropriate measurements on your computer screen. This is cheaper, quicker and safer than building actual circuits.

Capacitor – this is a device for storing relatively small amounts of electrical energy, similar to a battery; but able to charge and discharge in a very short

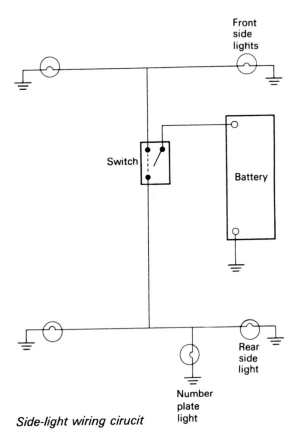

Front
side
lights

Switch

Battery

Rear
side
light

Number
plate
light

Side-light wiring cirucit

Figure 10.6 Vehicle side lights circuit in parallel.

amount of time, almost instantaneously. They are used in radios, TVs and amplifiers, and similar devices to enable signal tuning. Large-capacity ones are used in power supplies to machines where a constant smooth power supply is needed.

The charge stored in a capacitor is referred to as Q.

Charge Q = current (amps) × time (seconds)

$Q = I \times t$

Capacitance is the amount of charge (Q) that a capacitor can store for a given voltage, this is expressed in farads (F).

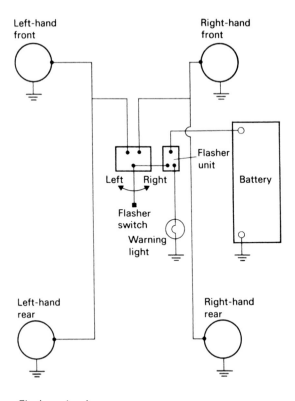

Flasher circuit

Figure 10.7 Vehicle indicator circuit in parallel.

Capacitance = charge/voltage

C = Q/V

Capacitors are usually marked with their capacity in farads (F). A farad is a rather large quantity, so usual values are in microfarads, that is 1×10^{-6} usually with the symbol µF.

Variable capacitor – used for tuning radios in to stations on different frequencies, now being superseded with solid-state devices.

Magnetism

Magnetism can be created in two ways, either using a permanent magnet or by using an electric current.

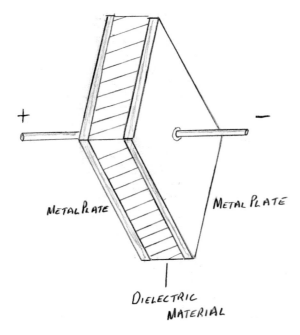

Figure 10.8 Capacitor.

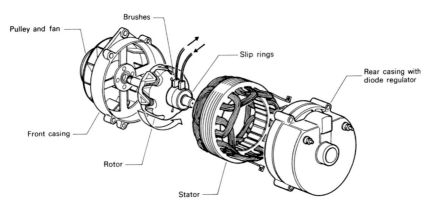

The components of the alternator

Figure 10.9 An alternator which produces alternating current which must be converted to DC to be used in a vehicle.

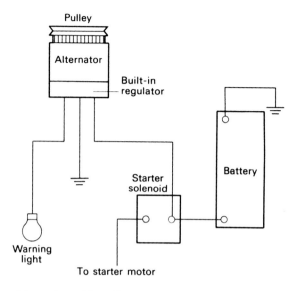

Alternator wiring diagram

Figure 10.10 Wiring diagram for alternator.

When an electric current is passed through a conductor it will generate a magnetic field; conversely when a magnetic field passes over a conductor it will generate an electric current. These are the basis of operation of an electric motor and a generator, respectively.

Faraday's law – Michael Faraday, a 19th-century English scientist, developed two laws related to magnets:

1. Whenever the magnetic field surrounding a coil changes it will cause an EMF (electricity) to be induced into the coil.
2. The amount of EMF induced is directly related to the rate of change of magnetic flux through the circuit.

Lenz's law – Emil Lenz, a 19th-century Russian/German scientist, carried out experiments to show that:

•   The direction of an induced EMF is always such that it opposes the force, or change, that is creating it.

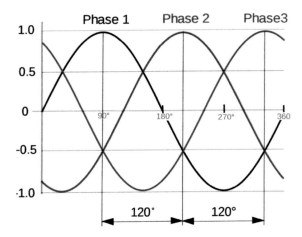

Figure 10.11 Three phases at 120° intervals of a three-phase power supply.

Alternating Current

Power stations, no matter what energy form they use – coal, other carbon-based fuel, nuclear or wind power – produce alternating current as it is easier and cheaper. The rotating armature within a coil naturally produces a sinusoidal wave pattern. The same principle applies to small petrol/diesel generators that are used in off-grid locations.

Solar panels produce DC current, which must then be converted into AC using an inverter for domestic or industrial use.

Tech note

Off-grid refers to both a remote location where electricity, and/or piped gas are not available; and where people choose to be independent of grid supply for personal life-style reasons.

Generator voltages – power stations may generate electricity at 25 kV, this may be supplied to the national grid at up to 400 kV. In local networks it is usually transmitted at 11 kV and then reduced in local sub-stations to either 415 V three-phase, or normal 230–240 V domestic AC.

Three-phase is used for high-voltage transmission as it is both cheaper and less subject to power loss. The higher the voltage, the lower the current (amperage) for any given amount of electrical power.

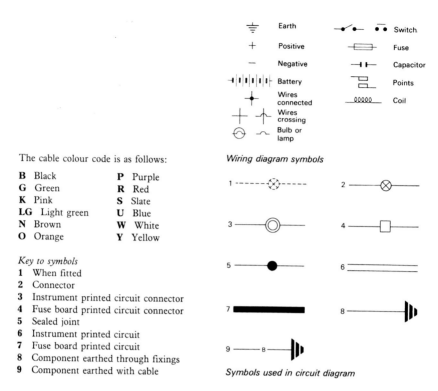

The cable colour code is as follows:

B	Black	**P**	Purple
G	Green	**R**	Red
K	Pink	**S**	Slate
LG	Light green	**U**	Blue
N	Brown	**W**	White
O	Orange	**Y**	Yellow

Key to symbols
1 When fitted
2 Connector
3 Instrument printed circuit connector
4 Fuse board printed circuit connector
5 Sealed joint
6 Instrument printed circuit
7 Fuse board printed circuit
8 Component earthed through fixings
9 Component earthed with cable

Figure 10.12 Circuit symbols.

Electric Cars

There are several different types of electric cars, motorcycles and vans:

• Pure electric vehicles (EV) these work by charging their batteries up from the mains supply.
• Hybrid electric vehicles (HEV) these have both electric power and an internal combustion engine – petrol or diesel – the engine may charge the batteries.
• Plug-in hybrid electric vehicles (PHEV) a combination of electricity and petrol or diesel with the advantage of being able to plug them in.
• Mild hybrid electric vehicles (MHEV) a mixture of internal combustion engine and electricity.

Tech note

The reader should be aware that new methodologies and processes are being introduced constantly for electric vehicles.

Important Points

- Electric vehicles operate at high voltages – currently between 100 and 1,000 volts.
- Never tow an electric vehicle – always use a transporter.
- Do not attempt to work on an electric vehicle unless you are fully trained for that vehicle.
- Following an accident keep clear of the vehicle, there is a risk of both electrocution and explosion.

Fuel cells, hydrogen – an increasing number of cars, vans and trucks are being run off hydrogen which is changed into electricity in fuel cells, in turn this feeds into electric motors to power the vehicle.

Tech note

Hydrogen is highly explosive and burns at about 2,600°C, which is hotter than most other fuels.

Figure 10.13 Race car instruments.

Figure 10.14 Car door speakers.

Skills and Questions

1. Find an example of the heating effect of electricity and sketch the electrical circuit for it.
2. Using a simple 1.5-V cell, some wire, a small magnet and a pencil, build a simple electric motor and/or a generator.
3. Using a solution of copper sulphate in a glass beaker, a 12-volt power supply, some wire and a piece of metal, copper plate the pieces of metal. Annotate which is the anode and which is the cathode.
4. Investigate the operation/layout of an electric vehicle of your choice.
5. Carry out experiments on both parallel and series circuits in your college's or training provider's workshop.
6. If possible, go on a field trip to an electricity-generating plant.
7. Place a magnet underneath a sheet of paper, sprinkle the paper with iron filings, shake the paper gently, can you see the lines of magnetic force.
8. Wind thin wire around a cardboard tube, connect the ends of the wire to a sensitive volt meter – galvanometer – move the magnet in and out of the tube, and note the meter readings.

Chapter 11

Standards Used in Engineering

Standards are documents setting out methods and procedures which may be followed by businesses, if these methods and procedures are followed that business, such as an engineering company, may be accredited by the standards organisation. The internationally recognised organisation is the **International Standards Organisation,** known by the abbreviation **ISO.** There are variations of these standards used in other countries for similar purposes and accredited by the standards organisation of that country. The use of these standards will be shown by the business using them on their documents and website. This is to show customers, and other organisations, that the procedures are being followed to ensure that the items produced are what they say they are. In other words, guaranteeing quality.

> **Tech note**
>
> Quality is a nebular word; it means different things in different contexts. In the context of standards, it is best seen as doing things in a set and approved manner.

In the UK, there is the British Standards Institute (**BSI**), in Europe there is the European Norm (**EN**), in America, the American National Standards Institute (**ANSI**) and Germany has the Deutsch Industrial Norm (**DIN**).

The common standards, which are identified by their number, and used in engineering companies across the world are as follows.

ISO 9001 – this is a quality management system widely used in many industries. One of its uses in engineering is that an engineered component can be tracked from conception to final usage. That is, the original design drawings will be signed and numbered, the source of the metal, or other material, will be recorded, the engineer and the machines used will be recorded and who it is sold to will be shown. This system is very useful as it shows the integrity of the item, and tracks it through its life journey.

DOI: 10.1201/9781003284833-11

An example of its importance is that should a part fail on an aircraft for instance, its history can be traced back, and the users of the other parts from the same batch informed of the possible failure so that action can be taken if needed.

Having this kind of information, being accredited to this standard, shows to customers and other commercial partners, that the company is reliable and concerned about the quality of its products, in other words, a good company to deal with making good-quality products or offering good reliable services.

ISO 45001 – this standard specifies requirements for an occupational health and safety (**OH&S**) management system, and gives guidance for its use, to enable organisations to provide safe and healthy workplaces by preventing work-related injury and ill health, as well as by proactively improving its OH&S performance. The concept behind this standard is the involvement of the workers in actually helping to manage the Health and Safety in the business. That is, the engineers and others contributing to the activities to prevent accidents. Those doing the job are best placed to say what is, and what isn't, safe. This involves recording meetings, planning actions and keeping accurate records. It promotes safety in the workplace.

ISO 14001 – this standard is about environmental management systems – **EMS**. It sets out a framework that encourages sustainability by undertaking a systematic approach to dealing with activities which affect the environment. It involves dealing with new legislation, and protecting the environment by responding to changing environmental conditions and at the same time considering both social and economic needs. If environmental management is carried out correctly costs should be reduced and workers' health improved; this should in turn reduce costs and improve profitability.

ISO 27001 – this standard is about information security management systems – **ISMS**. It sets out standards for managing the security of assets such as financial information, intellectual property (patents), employee information and other third-party information.

ISO 128 – this is about international drawing standards – it sets out conventions to make drawings readily readable throughout the world. See Chapter 3 on engineering drawing.

ISO 2768 – this is about geometric tolerancing, it is a way of making the drawing of complex engineered items simpler to understand.

Skills and Questions

1. In your college or workplace find copies of each of the standards, study them and see if you can identify which ones are used the most.

Chapter 12

Nature and Application of Standard Operating Procedures

- Give the safest way to do a job.
- Give the most cost-effective, if not the cheapest way to do a job.
- Be the most environmentally friendly and least wasteful way of carrying out a task.
- Comply with Health and Safety laws.
- Meet controlling body regulations.
- Be the best practice.
- Comply with warranty and other use conditions.
- Protect the public from danger.

Standard Operating Procedures – usually referred to by the abbreviation SOPs – are used to:

- Give the safest way to do a job.
- Give the most cost-effective, if not the cheapest way to do a job.
- Be the most environmentally friendly and least wasteful way of carrying out a task.
- Comply with Health and Safety laws.
- Meet controlling body regulations.
- Be the best practice.
- Comply with warranty and other use conditions.
- Protect the public from danger.

Tech note

In some organisations the word standard is replaced with the word standing.

DOI: 10.1201/9781003284833-12

SOPs may include:

- Procedures – detailed step-by-step instructions.
- Work method statements – also called safe working methods.
- Health and safety details – precise details of compliance with specific rules and regulations.
- Code of conduct – list of rules of behaviours including dress codes and PPE details.
- Administrative protocols – particularly related to IT systems, both hardware and software.
- Quality control systems – compliance with quality control often means the generation of photographic evidence along with inspectors' signatures.
- Good practice – abbreviation is GxP –may cover the use of particular tools and equipment.
- Run book – also referred to as a diary or a log. This is especially important when updating software or other computer-controlled equipment.
- Tools list – this will include specific hand tools and special tools as well as diagnostic equipment.

In many installation and repair industries there are workshop manuals which state the SOP for every job. For automotive repair and servicing the workshop manual states how to fault find, and how to rectify, or replace components. Once complete the workshop supervisor will check the repair and road test the vehicle. On aeroplanes and trains the same manuals exist; but in much more detail. The supervisor will check each stage in detail and sign the document to show clearly by whom, when and where the repair was carried out.

Tech note

A commuter train may be carrying over 1000 passengers, the incorrect tightening of one bolt, not following the SOP, could cause an accident de-railing the whole train and lead to several fatalities. This could lead to a custodial sentence for all those involved.

During your engineering studies you are very likely to need to both work to, and to write, an SOP. There are many variation on SOPs, the following is offered as an exemplar, which can be used as a proforma.

Removing Vehicle Engine

Make:

Model:

Date:

Technician:

Tools required: M spanner set, M socket set, vehicle hoist, engine hoist, hoist to engine adaptor.

Task No	Activity	Special points	Technician notes, confirm with tick
1	Read workshop manual appropriate pages	Mark any special points	
2	Inspect vehicle for damage, report as appropriate	Report any problems immediately	
3	Check risk assessment documents, carry out further assessments if appropriate – is it safe to do this job on this vehicle?	Take action as required – report to workshop manager. Note COSHH data	
4	Confirm that appropriate PPE is in place along with protective covers as needed	List PPE and protective covers	
5	Disconnect battery	Disconnect earth terminal first	
6	Drain coolant	Store	
7	Drain engine oil	Ensure it goes into reclaim tank	
8	Drain and evacuate air conditioning fluid	Check that the pump receiver tank has sufficient capacity before connecting	
9	Confirm positive status for all above points		
10	Check mounting of vehicle on hoist is secure and raise to required height	Check that all mounting points are in place and secure	
11	Undo fastenings which are accessible from underneath	Support loose end exhaust system	

Removing Vehicle Engine Make: Model: Date: Technician:			
12	Lower vehicle		
13	Secure engine lifting hoist as shown in workshop manual		
14	Remove upper level engine fastenings	At this point the lifting hoist should be taking the strain of the engine weight	
15	Use hoist controls to lift engine out of engine bay and transfer to work bench		
16	Carry out inspection of work area and clear/clean as needed	Ensure all fastenings and unattached parts are correctly labelled and stored	
17	Confirm that the working area is now clean and safe		

Skills and Questions

1. Investigate the SOPs which are available either at your college, or your place of work, make notes of points which are important to you.
2. Write an SOP for a job which you are planning, ensure that a risk assessment is carried out and where appropriate COSHH data.
3. Write up a report on a job that you are working on, take photographs of each step so that you can describe this to a colleague or your tutor. Often these reports are needed as part of your college work.

Chapter 13

Health and Safety – Standards, Acts, Legislation and Risk Analysis

This chapter follows on from Chapter 2, which is an introduction to health and safety. In this chapter we will take a higher level approach. When you complete your T Level, or other engineering Level 3 qualification, you will be at a level where your understanding and skill are at an average level for the engineering industry. In other words, you will be at a level at which you will be not only expected by your employer to take responsibility for your own and other's health and safety, but to take a lead in this area by actively carrying out health and safety tasks and being involved in the decision-making discussions.

> **HSE website** – Health and Safety standards, acts and legislation related to Statutory Instruments – also called Regulations – are constantly changing. The student and teacher are both encouraged to confirm the latest situation by checking on the Health and Safety Executive (HSE) website. The number of Acts of Parliament and Statutory Instruments is many pages long, the documents themselves will fill several library shelves. The legislation in the UK, which follows on from the Health and Safety at Work Act 1974, is more or less identical to that of other European countries, and most other, non-European countries, base their legislation on that of the UK too.

The following sections discuss some of the more salient points, ones which you are likely to be examined on.

The six tenets of the Health and Safety at Work Act 1974 are:

1. **Provide a safe place of work**
   This covers the physical workplace to ensure that premises are up to standard. It includes considerations like fire safety, cleanliness, waste

DOI: 10.1201/9781003284833-13

management and the handling of harmful substances. Workplaces such as construction sites or medical labs will have more factors to consider than an office building.

2. **Provide safe equipment**
   Any equipment that's used at work, including computers and electronic devices, needs to be maintained to ensure it's safe to use. This would usually involve periodic safety checks by an appointed person and a set process to report any faults so they can be repaired.

3. **Ensure staff are properly trained**
   In order to foster a safe workplace, it's important to surround yourself with responsible and competent staff. This includes both on-site training (to ensure all staff have been trained to use equipment properly) and general health and safety training, such as manual handling and fire safety.

4. **Carry out risk assessments**
   Thorough risk assessments are an important part of HASAWA so employers can put in place appropriate preventative actions for each risk identified. One caveat to the legislation is that it requires employers to protect 'as far as is reasonably practical' the health and safety of their employees. The key thing here is risk. If something is identified as a very low-risk factor, and would incur significant cost to mitigate, it might not need to be actioned.

5. **Provide proper facilities**
   This covers the basics like toilets, clean drinking water, heating and air conditioning. If you provide kitchen facilities, any appliances will need to be checked and maintained accordingly. Employees also have their part to play in keeping these areas clean and tidy and in taking responsibility for their own health and safety.

6. **Appoint a competent person to oversee health and safety**
   This would be a dedicated person who would be responsible for ensuring that all health and safety duties are being carried out and adhered to by employees. This might include routine safety inspections, managing day-to-day operations and working with safety reps throughout the business. If your organisation is attached to a union, you'll also be required to liaise with an appointed representative.

Health and Safety at Work Act 1974

The Health and Safety at Work Act 1974 is the primary piece of legislation covering occupational health and safety in Great Britain. It's sometimes referred to as HSWA, the HSW Act, the 1974 Act or HASAWA.

> **Duty of Care** – under common law all employers and employees have a duty of care to behave in a safe and responsible manner. Failure to do this, referred to as **negligence**, may lead to both **civil** and **criminal** legal actions. This may be in addition to legal actions under the HSAWA.

Working at Heights – the greatest number of deaths at work are caused by falling from heights, this is about 30% of all deaths. Many serious falls are from a height of less than 2 metres. Serious injuries from falls must be reported to the HSE.

Many engineering jobs entail working at heights, the power industry and telecommunications are obvious ones, installation of mechanical systems such as air-conditioning and engineering construction of many industrial and commercial buildings employs lots of engineers. Working at heights also includes working at a level where there is a pit or chance to fall to a lower level; examples include pits and hoists in motor vehicle garages, and cellars and evacuation work.

Before working at height you must work through these simple steps:

- Avoid work at height where it is reasonably practicable to do so.
- Where work at height cannot be avoided, prevent falls using either an existing place of work that is already safe or the right type of equipment.
- Minimise the distance and consequences of a fall, by using the right type of equipment.
- Do as much work as possible from the ground.
- Ensure workers can get safely to and from where they work at height.
- Ensure equipment is suitable, stable and strong enough for the job, maintained.
- Make sure you don't overload or overreach when working at height.
- Take precautions when working on or near fragile surfaces.
- Provide protection from falling objects.
- Consider your emergency evacuation and rescue procedures.

The definition of safety equipment is not prescriptive; but the use of safety harnesses, head protection, and podiums or lifting platforms is the usual industry norm.

Electricity at Work – About 10% of Deaths at Work are Caused by Electrocution

The Electricity at Work Regulations require regular inspection of electrical systems and circuits in all factories and workshops.

The purpose of these regulations is to prevent death or personal injury to any person from electrical causes in connection with work activities.

'Injury' means death or injury to any person from:

- Electric shock
- Electric burn
- Fires of electrical origin
- Electric arcing
- Explosions initiated or caused by electricity.

The difficulty is that electricity cannot be seen, nor smelt, and even relatively low voltages can kill. It is known that small 9-volt batteries have caused death.

Electrical faults often cause fires. They can cause shocks, burns, flashes and fire.

Before working on electrical circuits you must ensure that they are isolated, not just turned off. Isolation means that all the electrical cables are disconnected from the power supply, using some form of isolator box. Isolators tend to be built into fuse boxes. If you are servicing or repairing machine tools, they usually have an isolator switch which can be locked with a key. Only qualified personnel are allowed to work on electrical circuits and equipment.

Before using portable electrical equipment, such as an electric drill, you should check that it has been PAT tested and you must visually inspect it for damage, including the full length of the cable and the plug. If in any doubt seek advice.

Tech note

PAT – Portable Appliance Testing – this comprises of: visual inspection, and meter reading tests for earth continuity, insulation resistance and lead polarity. It is usually referred to as PAT testing.

Use low-voltage equipment where possible, preferably cordless, for both safety and ease of use.

Electric cars may have battery voltages in excess of 1,000 volts. Their batteries store large amounts of power, sometimes exceeding 100 kW. To put this into perspective, this would be enough to supply electricity to a row of houses.

Safety note

Only work on electric vehicles, that is cars, motorcycles, electric cycles and scooters if you have been trained to do this.

Manual Handling – about 30% of injuries reported at work are related to manual handling situations. All employees should be aware of the correct procedures. Injuries can be sustained moving even a light object if the correct procedure is not followed. Before carrying out a manual handling activity you should carry out a dynamic risk assessment, this will probably include:

- Considering the task to be done.
- Determining the load to be moved.
- Being aware of your carrying capacity and the equipment which is available.
- Being cognisant of the working environment.

Tech note

A dynamic risk assessment is done in your head before attempting a new activity, simply look and think what needs to be done, if you spot a hazard, stop and seek advice.

RIDDOR

Deaths and injuries

If someone has died or has been injured because of a work-related accident this may have to be reported. Not all accidents need to be reported, other than for certain gas incidents, a RIDDOR report is required only when:

The accident is work-related, and it results in an injury of a type which is reportable.

Types of reportable injury

The death of any person

All deaths to workers and non-workers, with the exception of suicides, must be reported if they arise from a work-related accident, including an act of physical violence to a worker.

Specified Injuries to Workers

The list of 'specified injuries' in RIDDOR 2013 replaces the previous list of 'major injuries' in RIDDOR 1995. Specified injuries are (regulation 4):

- fractures, other than to fingers, thumbs and toes
- amputations
- any injury likely to lead to permanent loss of sight or reduction in sight
- any crush injury to the head or torso causing damage to the brain or internal organs
- serious burns (including scalding) which:
  - covers more than 10% of the body
  - causes significant damage to the eyes, respiratory system or other vital organs
- any scalping requiring hospital treatment
- any loss of consciousness caused by head injury or asphyxia
- any other injury arising from working in an enclosed space which:
- leads to hypothermia or heat-induced illness
- requires resuscitation or admittance to hospital for more than 24 hours.

Over-Seven-Day Incapacitation of a Worker

Accidents must be reported where they result in an employee or self-employed person being away from work, or unable to perform their normal work duties, for more than seven consecutive days as the result of their injury. This seven-day period does not include the day of the accident, but does include weekends and rest days. The report must be made within 15 days of the accident.

Over-Three-Day Incapacitation

Accidents must be recorded, but not reported, where they result in a worker being incapacitated for more than three consecutive days. If you are an employer, who must keep an accident book under the Social Security (Claims and Payments) Regulations 1979, that record will be sufficient.

Non-Fatal Accidents to Non-Workers (e.g., Members of the Public)

Accidents to members of the public or others who are not at work must be reported if they result in an injury and the person is taken directly from the scene of the accident to hospital for treatment to that injury. Examinations and diagnostic tests do not constitute 'treatment' in such circumstances.

There is no need to report incidents where people are taken to hospital purely as a precaution when no injury is apparent.

If the accident occurred at a hospital, the report only needs to be made if the injury is a 'specified injury' (see above).

Occupational Diseases

Employers and self-employed people must report diagnoses of certain occupational diseases, where these are likely to have been caused or made worse by their work: These diseases include (regulations 8 and 9):

- carpal tunnel syndrome;
- severe cramp of the hand or forearm;
- occupational dermatitis;
- hand-arm vibration syndrome;
- occupational asthma;
- tendonitis or tenosynovitis of the hand or forearm;
- any occupational cancer;
- any disease attributed to an occupational exposure to a biological agent.

Dangerous Occurrences

Dangerous occurrences are certain, specified near-miss events. Not all such events require reporting. There are 27 categories of dangerous occurrences that are relevant to most workplaces, for example:

- the collapse, overturning or failure of load-bearing parts of lifts and lifting equipment;
- plant or equipment coming into contact with overhead power lines;
- the accidental release of any substance which could cause injury to any person.

Additional categories of dangerous occurrences apply to mines, quarries, offshore workplaces and relevant transport systems (railways, etc.).

Gas Incidents

Distributors, fillers, importers and suppliers of flammable gas must report incidents where someone has died, lost consciousness or been taken to hospital for treatment to an injury arising in connection with that gas. Such incidents should be reported using the Report of a Flammable Gas Incident – an online form.

Registered gas engineers (under the Gas Safe Register) must provide details of any gas appliances or fittings that they consider to be dangerous, to such an extent that people could die, lose consciousness or require hospital treatment. The danger could be due to the design, construction, installation, modification or servicing of that appliance or fitting, which could cause:

- an accidental leakage of gas;
- incomplete combustion of gas or;
- inadequate removal of products of the combustion of gas.

Unsafe gas appliances and fittings should be reported using the Report of a Dangerous Gas Fitting – an online form.

Mechanical – about 14% of deaths at work are caused by contact with machinery. All moving parts must be fitted with guards. The guards should have switches fitted so that the machine cannot be operated without the guard in place. Most modern machinery is fully enclosed with tamper-proof switching.

When operating any machinery the correct PPE must be worn. This usually entails close-fitting clothing and safety spectacles. No loose clothing, jewellery, watches or lanyards should be worn.

Automotive – about 25% of deaths at work are caused by being struck by a moving vehicle. Speed is often a contributor to this situation. Working areas are advised to limit speed to walking pace, which is approximately 4 mph. The use of bleepers and flashing lights is normal practice. The growth of autonomous vehicles is an issue and requires an operating protocol within this environment.

Noise – The level at which employers must provide hearing protection and hearing protection zones is 85 dB(A) (daily or weekly average exposure) and the level at which employers must assess the risk to workers' health and provide them with information and training is 80 dB(A).

COSHH – Control of Substances Hazardous to Health

This includes:

- dusts from mechanical cutting, shaping and abrasive blasting;
- gases and fumes from welding, soldering and cutting;
- mists and germs in metal-working fluids;
- lubricants, adhesives, paints, degreasing and stripping fluids;
- plating and pickling fluids, and molten salt baths;
- fluid treatment products.

Control measures include:

- dust, fume or vapour extraction;
- respirators;

- fluid maintenance;
- skin checks.

Employers may need to use health surveillance, that is, checking employees' health for any adverse effects related to work; this involves checking skin for dermatitis or asking questions about breathing and may need to done by a doctor or nurse.

Employers provide equipment to protect the workers' health, such as:

- extraction of dust, fumes or mist;
- control of bacteria in cutting fluids;
- personal protective equipment.

Employees have a duty to use these properly and co-operate with any monitoring and health surveillance.

Employers should have readily available a database of information about any potentially hazardous substances in use.

Risk Assessment

Before carrying out any task you must complete a risk assessment. In engineering, where tasks are usually done to a set pattern, that is using a standard operating procedure – SOP – it is usual to have completed risk assessment forms readily available.

Risk Assessment Form

The usual format of a risk assessment form asks questions about the task to be completed and the controls in place to either illuminate, or mitigate, the risk.

Hierarchy of Control

Taking an overview of the process, or task, it is usual to refer to the hierarchy of control. The move to fully enclosed and fully automated machine tools has almost eliminated any contact with rotating machinery.

Risk and Hazard Matrix

It is normal to carry out a risk assessment, and then consider the likelihood of this happening. The matrix helps inform decisions about how to tackle a task.

Cost of Accidents

All accidents cost money, some of the costs which may arise after the accident may stretch into millions of pounds. These costs may this include the following:

- Immediate cost of dealing with the situation – first aid, loss of actual work materials, hospital or other treatment, loss of workers' time, loss of completed item, administration costs.
- Investigation of accident: loss of time that the job is stopped, costs for inspectors and consultants, managers' time.
- Business costs: loss of contract, loss of future business, increase in insurance costs, dealing with the aftermath of the accident, operational delays.
- Legal costs: solicitor and barrister fees, fines and sanctions, civil and criminal proceedings.

Skills and Questions

Care for Health and Safety, and the environment is something which MUST be part of EVERY job that you do EVERY day. However, these skills need developing, and do not always come easily – keep practising until Health and Safety are second nature for you. Meanwhile try these questions:

1. Look at the HSE website and state six items that are appropriate to the branch of engineering which interests you the most.
2. Write out a risk assessment for a job which you are going to, or already working on.
3. Sketch a plan drawing of the workshop which you work in and show the various safety features – such as stop buttons and fire extinguisher locations.
4. Make a poster showing how dermatitis can be prevented in the engineering environment.
5. Discuss with colleagues the steps needed to ensure safe lone working.
6. List three occupational diseases which must be notified to the HSE if contracted at work.
7. Name three serious injuries which must be notified to the HSE if they happen at work.
8. Obtain a COSHH data sheet for a substance which you use at work.
9. State the items which may be included in the cost of an accident at work.
10. Give an example of a permit to work.

Chapter 14

Business, Commercial and Financial Awareness

The basis of business and commerce is usually to make money. It may be not for profit, that is, it maybe for charitable purposes, or to provide an essential service. No matter which way, it is likely to have both income and expenditure. The income needs to be equal to, or greater than, the expenditure for the business to remain viable. Manufacturing engineering firms usually create wealth in two ways. They make objects which are valuable in themselves, the ownership of these objects adds to the wealth of whoever possesses them. The objects have been sold by the maker for more than they cost to make, so making the manufacturers a profit.

Tech note

Definitions to remember: business refers to a person's, or a company's, regular occupation, profession or trade, such as making cars. Commerce is about buying and selling objects.

SME – Small–Medium Enterprise, 99% of businesses in the UK are SMEs, these are companies employing less than 250 employees. Worldwide the percentage is 90%. Frequently, SME engineering companies in the manufacturing and service sectors supply good and services to larger corporations.

Added value – a piece of metal may cost £100, if it is machined into a useful object then it may sell for £200. It is still the same piece of metal; but its value has now risen. In the same way a broken tractor will only have scrap value, if an engineer repairs it, the tractor will have a new value as it can now be used. Again, this is added value.

Wealth generation – engineering generates wealth. An example of this is building a bridge. The engineering company building the bridge will be paid for the materials and labour, thereby making a profit – wealth. The owners of the bridge may charge people for using the bridge, again generating an income – wealth. Those using the bridge, although paying a charge, will

DOI: 10.1201/9781003284833-14

save time and perhaps travel a shorter distance, so that they may do more work – generating more wealth.

Types of Companies

Any company is known as a legal entity, that is, it has both legal rights and legal obligations. These rights and obligations depend on the way that the company is drawn up, that is, how it is legally formed and its constitution. All companies operating in the UK must be registered with **Companies House**. This is, in reality, a government department. Companies House keeps a record of all company names and financial returns for taxation purposes. All companies, except sole traders, must file annual reports to Companies House. Sole traders make reports through their normal individual annual tax returns.

Sole trader – a very small company, there are a large number of small businesses in the various engineering trades and professions operating as sole traders. The sole trader may trade with a registered business name; but the sole trader is fully personally responsible for all debts and charges. This means that the sole trader could lose all personal possessions in the case of an error, or claim.

Set up as a sole trader – If you're a sole trader, you run your own business as an individual and are self-employed. You can keep all your business's profits after you've paid tax on them. You're personally responsible for any losses your business makes. You must also follow certain rules on running and naming your business.

When you need to set up as a sole trader – You need to set up as a sole trader if any of the following apply:

- You earned more than £1,000 from self-employment between 6 April in you first year and 5 April the following year.
- You need to prove you're self-employed, for example, to claim Tax-Free Childcare.
- You want to make voluntary Class 2 National Insurance payments to help you qualify for benefits.

How to Set Up as a Sole Trader

To set up as a sole trader, you need to tell HMRC that you pay tax through Self-Assessment. You'll need to file a tax return every year.
You'll need to:

- Keep business records and records of expenses.
- Send a Self-Assessment tax return every year.

- Pay Income Tax on your profits and Class 2 and Class 4 National Insurance – use HMRC's calculator to help you budget for this.

Limited Company – a limited company limits its liability when it is set up to a given amount of shares. Each of the directors will possess a number of shares.

A limited company is a company **'limited by shares'** or **'limited by guarantee'**.

Limited by shares companies are usually businesses that make a profit. This means the company:

- is legally separate from the people who run it
- has separate finances from your personal ones
- has shares and shareholders
- can keep any profits it makes after paying tax.

Limited by guarantee companies are usually 'not for profit'. This means the company:

- is legally separate from the people who run it
- has separate finances from your personal ones
- has guarantors and a 'guaranteed amount'
- invests profits it makes back into the company.

As a director of a limited company, you must:

- follow the company's rules, shown in its articles of association
- keep company records and report changes
- file your accounts and your Company Tax Return
- tell other shareholders if you might personally benefit from a transaction the company makes
- pay Corporation Tax.

PLC – Public Limited Company, a PLC must have an issued share capital of at least £50,000, of which at least 25% of the nominal value and the whole of any share premium must be paid up when the company is registered. A PLC must have at least two directors. A PLC must have at least two shareholders. A PLC must have a company secretary.

CIC – Community Interest Company, a type of social enterprise, usually for charitable purposes. These are similar to registered charities, usually being run for a specific purpose, for a specific section of the community.

Partnership – when two or more people work together to share profits and expenses it is called a partnership. The rules are similar to those of a sole trader.

LLP – Limited Liability Partnership, this is similar to a limited company in that the liability is limited to the separate entity. This type of business is usually used by solicitors and consultants.

Company Structure

Medium and larger companies tend to be hierarchical in structure. That is, they will have a managing director, may be assisted by a few other directors, and these people may own the company. There will be a number of heads of departments who report to the directors. The heads of the departments will be in charge of the staff within their departments.

Small companies tend to have a very flat structure, where the company directors work in the workshop – or shop floor as it is often called – alongside the other employees.

Increasingly, companies are organised into teams, each team having special skills for particular jobs. An example is to have a sales team, a manufacturing team and an installation team. The members of each team will work to achieve set objectives in a given time period.

Communications – both inside any company, and to those outside, that is to customers, or suppliers, communications are very important. With the increasing emphasis on GDPR and equal opportunities legislation, many companies have set formats for most communications, that is letters and emails.

Profit and Loss Account (PLA)

This contains essential data to see if the company is actually making a profit or a loss. With the growth of Community Interest Companies (CICs), generally not-for-profit organisations, these need to be crystal clear to show all incomes and outgoings. This is essential to assure HMRC and others where the money is going.

Any Engineering Company	
Profit and Loss Account	
Income	£
Component sales*	25,000
Accessory sales*	4,000
Service work	3,000
Repair work	10,000

*less purchase costs

	£
Total	42,000

Expenses	£
Rent of premises	5,000
Business rates	1,000
Electricity	250
Water	300
Gas	300
Transport	1,500
Technology	250
Total	8,600
Trading Profit	£33,400

Balance Sheet

The balance sheet sets out the company's position with the relationship between what it owns and what it owes at any one point in time; to keep abreast of trading it is normal to produce a balance sheet for each quarter of the year and compare it to previous similar quarters. The bicycle industry has variation with the seasons, so each quarter's results may be very different; but the comparison of any quarter with those of the previous year's same quarter should give a good indication of the trading situation.

1
Any Engineering Company
Balance Sheet

	Last Year	This Year
	£	£
Current Assets		
Cash in hand	1500	500
Cash in bank	10,500	13,000
Stock at valuation	120,000	130,000
30-day assets	20,000	15,000

Inventory assets	10,000	12,000
Subtotal	**16,200**	**170,500**
Long-term assets		
Tools and equipment	10,000	7000
Goodwill	2000	5000
Premises	30,000	40,000
Subtotal	**42,000**	**52,000**
<u>Total current assets</u>	<u>204,000</u>	<u>225,000</u>
Current liabilities		
Bills payable	30,000	32,000
Loans payable	10,000	8000
Tax payable	10,000	10,000
<u>**Total current liabilities**</u>	<u>**50,000**</u>	<u>**50,000**</u>

Acid-test ratio (ATR) – this is the absolute test of a company's viability. It is calculated by dividing the company's total current assets (TCA) by the total current liabilities (TCL).

ATR = TCA/TCL

For the two years shown in the balance sheet these are:

2019 ATR = 204,000/50,000 = 4.08

2020 ATR = 222,500/50,000 = 4.45

It can be seen that the ATR has improved as the company has matured. If the ATR is greater than unity (1) the company is viable; if less then unity there is likely to be a need to borrow money.

Sales – this is the total income for goods and services provided.

Cost of sales – this is the total cost of making the goods or providing the services, it includes materials, wages, energy, advertising and other general expenses.

Tax and VAT – in all countries tax is payable on profits. Some countries have lower tax rates than others. It is worked out as a rate (percentage) of the difference between income (sales) and expenses (cost of sales). There are usually different rates (percentages) for different amounts. There are also different rates for the different types of companies.

Value Added Tax, usually just referred to as VAT, is a tax on the sale of the goods, the amount payable is typically 20% of the difference between the cost of the goods and the sale price of the goods – that is the amount of value added, hence its name.

External factors – may affect a business for the better or worse. Recent wars and the COVID pandemic are examples that could not be foreseen nor ignored. Many engineering companies changed what they were manufacturing to help alleviate the situations. Looking back in history, engineers have always risen to the challenge of supplying what was needed.

Economic environment – the economy of the country also affects engineering production. In good economic times, expensive cars and high-quality electronic goods are in demand, amongst other items. When the economy is weak, and unemployment growing, the need is for utility cars and basic essential items.

Location – where the business is situated is very important. At the start of the Industrial Revolution cities in the middle of England boomed as they were close to the iron ore and coal needed to make steel. This is now less important; but cities in the Midlands remain important centres for manufacturing as they have a high population of skilled people.

Costing and pricing – costing in engineering means adding up what actually goes into the product, this may include:

- Materials
- Labour
- Machinery
- Tooling
- Capital depreciation
- Other overheads such as rent, rates, energy and other services.

When bringing a new product to market there are a lot of other costs which are not always clear to see, such as: design costs, new tooling, development and testing.

Pricing, how much you are going to sell your goods or services for, depends on your market, and how many of a particular product or service your company is going to sell.

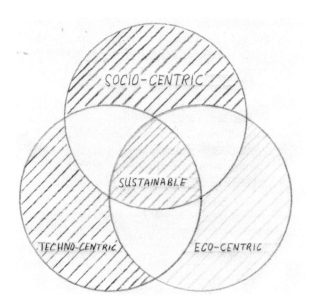

Figure 14.1 The three centrics: socio-centric, techno-centric and eco-centric. Where they overlap is a sustainable balance.

Sustainability – the three centrics – socio, techno and eco – need to overlap to maintain sustainability. To maintain any business it is necessary to maintain a constant stream of products for the market, and have products which are attractive to customers.

Laws – all businesses are subject to various international and national laws, as well as local rules and regulations. Many of these are covered in the Health and Safety chapters of this book. It is beyond this textbook to cover all the possible laws of the companies which you may be involved with, however GDPR is currently relatively new, and very important, so it is worth considering in detail.

GDPR – The Data Protection Act 2018 is the UK's implementation of the **General Data Protection Regulation**. Similar regulations operate in most countries.

Everyone responsible for using personal data has to follow strict rules called 'data protection principles'. They must make sure the information is:

- used fairly, lawfully and transparently
- used for specified, explicit purposes

- used in a way that is adequate, relevant and limited to only what is necessary
- accurate and, where necessary, kept up to date
- kept for no longer than is necessary
- handled in a way that ensures appropriate security, including protection against unlawful or unauthorised processing, access, loss, destruction or damage.

There is stronger legal protection for more sensitive information, such as:

- race
- ethnic background
- political opinions
- religious beliefs
- trade union membership
- genetics
- biometrics (where used for identification)
- health
- sex life or orientation.

There are separate safeguards for personal data relating to criminal convictions and offences.

Individuals' Rights

Under the Data Protection Act 2018, you have the right to find out what information the government and other organisations store about you. These include the right to:

- be informed about how your data are being used
- access personal data
- have incorrect data updated
- have data erased
- stop or restrict the processing of your data
- data portability (allowing you to get and reuse your data for different services)
- object to how your data are processed in certain circumstances.

You also have rights when an organisation is using your personal data for:

- automated decision-making processes (without human involvement)
- profiling, for example, to predict your behaviour or interests.

Skills and Questions

1. Draw a plan of your local area and indicate in it the engineering companies, noting what sort of engineering they do.
2. Go onto the Companies House website, it is freely available on the Internet, and make a note of the details of either a local company, or a company elsewhere, that interests you.
3. In the company that you are having work experience with, find the guidelines, or instructions, related to sending emails.
4. Investigate what opportunities are available for you to start a small business, or a big one if you have money to invest. It is usual for entrepreneurs to start businesses when students. Ask your parents/ careers, friends, neighbours, college staff for suggestions.
5. Why is sustainability important to any business?

Chapter 15

Professional Responsibilities, Attitudes and Behaviours

Engineering is a serious profession and requires a disciplined approach. Whether fully qualified, a student, or working as an apprentice, all those engaged in engineering are required to adhere to a positive professional code, this includes: professional procedures, attitudes and behaviours as set out in the Engineering Council and Royal Academy of Engineering document called *Engineering Ethics: Maintaining Society's Trust in the Engineering Profession.*

Engineering Council

As the regulatory body for the UK engineering profession, the Engineering Council sets and maintains internationally recognised standards of professional competence and commitment. These are detailed in the UK Standard for Professional Engineering Competence, known as UK-SPEC.

There are three registration titles:

- **Chartered Engineer (CEng)**
- **Incorporated Engineer (IEng)**
- **Engineering Technician (EngTech)**

These three different levels of registration relate to the level of qualifications and experience of the individual.

EngTech is related to Level 3 qualifications – such as T Levels, with suitable training and workshop experience.

IEng is related to Level 5/6 qualifications – such as a degree, with suitable supervisory, or project management, experience.

CEng is related to post-graduate qualifications – such as a MEng degree and a senior level of experience and responsibility.

The Engineering Council is the overarching body, that is, it gives the power to put forward registration to the individual institutions, which are often referred to as professional bodies, or learned societies. The list below

DOI: 10.1201/9781003284833-15

is of the institutions that are currently operating under the engineering regulations.

British Institute of Non-Destructive Testing (BINDT)

Chartered Association of Building Engineers (CABE)

Chartered Institution of Building Services Engineers (CIBSE)

Chartered Institution of Civil Engineering Surveyors (CICES)

Chartered Institution of Highways & Transportation (CIHT)

Chartered Institute of Plumbing and Heating Engineering (CIPHE)

Chartered Institution of Water and Environmental Management (CIWEM)

Energy Institute (EI)

Institution of Agricultural Engineers (IAgrE)

Institution of Civil Engineers (ICE)

Institution of Chemical Engineers (IChemE)

Institution of Engineering Designers (IED)

Institution of Engineering and Technology (IET)

Institute of Explosives Engineers (IExpE)

Institution of Fire Engineers (IFE)

Institution of Gas Engineers and Managers (IGEM)

Institute of Highway Engineers (IHE)

Institute of Healthcare Engineering and Estate Management (IHEEM)

Institution of Lighting Professionals (ILP)

Institute of Marine Engineering, Science & Technology (IMarEST)

Institution of Mechanical Engineers (IMechE)

Institute of Measurement and Control (InstMC)

Institution of Royal Engineers (InstRE)

Institute of Acoustics (IOA)

Institute of Materials, Minerals and Mining (IOM3)

Institute of Physics (IOP)

Institute of Physics and Engineering in Medicine (IPEM)

Institution of Railway Signal Engineers (IRSE)

Institution of Structural Engineers (IStructE)

Institute of Water

INCOSE UK, the UK Chapter of the International Council on Systems Engineering (INCOSE)

Permanent Way Institution (PWI)

Nuclear Institute (NI)

Royal Aeronautical Society (RAeS)

Royal Institution of Naval Architects (RINA)

Safety and Reliability Society (SaRS)

The Society of Operations Engineers (SOE)

The Welding Institute

Becoming Professionally Registered

The first step to becoming professionally registered with the Engineering Council as EngTech, IEng or CEng is membership of a licensed professional engineering institution. The institution will act as the awarding body for your registration.

When choosing which institution to contact it is best to join one closest to the discipline of engineering you work in. They will be best placed to assess your competence for professional registration. If you find that there are two or three suitable institutions, you might wish to join more than one, or contact all those suitable to choose which best meets your needs.

There are many benefits of professional engineering institution membership, including:

- Recognition of the member's professional status, which may include post-nominals
- Support and guidance with professional development, including the registration process
- Technical regional events, seminars, conferences, often at a special membership rate
- Monthly journals and other technical publications
- Access to technical library
- Careers advice
- Specialist interest groups and other networking opportunities
- Health and legal advice.

The Statement of Ethical Principles is underpinned by four fundamental principles:

1. Honesty and integrity
2. Respect for life, law, the environment and public good
3. Accuracy and rigour
4. Leadership and communication.

This can be summed up as good professional behaviour, just as you would expect from any other professional such as a doctor or a lawyer.

Questions and Skills

1. Research the various institutions which are of interest to you, choose one and find out what you need to do to join it.
2. If possible attend a local meeting – all institutions are keen to get new members.
3. If appropriate join as a student member – this is usually free.

Chapter 16

Stock and Asset Management

Engineering business, and most other businesses too, have some form of stock, that is, goods or raw materials which their product is based on, and completed, manufactured or repaired items which have not been sold, or paid for. Assets refers to the workshop, the tools and equipment and the people who work there.

Tech note

Stock – the goods, or items, that a company keeps in stock (storage) for use, consumption or sale.

Asset – something such as a piece of equipment, or someone (a person), that enables the business to operate.

Remember that both stock and assets have a monetary value to the business.

The management of both stock and assets is a distinct process consisting of planning, organising, actuating and controlling.

Planning – it is important to have a storage plan; a layout plan for the storage area, or warehouse. This is to ensure that items are easy to access and can be taken to the working area or delivered without moving other items.

Inventory control – this is an essential for looking after the stock. It is important to ensure that stock records are kept up to date and that there is sufficient stock to keep production continuous. The inventory will also highlight if there is:

- Any damage
- Pilferage
- Loss of items
- Lack of recording

DOI: 10.1201/9781003284833-16

Logistics

JIT – just in time is a system originally devised by Henry Ford to ensure that the parts for building a car arrived just in time for them to be fitted. Explaining this in more detail, the original system, still used by some car manufacturers, and used in many other engineering organisations, is to order parts in bulk and store them in a part department, or warehouse. For small parts this does not need much space; for the parts to make bigger items this may be a considerable amount of space. With JIT no storage space is needed. As the car, or other large item, progresses down the assembly line the required parts are delivered. The current way of ordering a new car is on-line, where the specification is chosen. The manufacturers order the parts to the required specification to arrive as the car is being built.

Stock control – looking after the stock is very important, ensuring that it is protected from the weather, damage and pilferage.

Stock taking – this is an activity which is usually carried out once a year to coincide with the end of the financial year. The process involves checking the stock against the inventory, noting missing or damaged items, to arrive at a final value for the stock.

Writing off/down – this refers to the value of stock items. It is usually carried out during the stock-taking period. It involves inspecting the stock and re-setting the value of each item. Often these items are sold off as scrap, or to be repurposed.

Stock life – some stock has an infinite life; other stock is time limited. Some materials deteriorate quickly. Metal in stock will rust or corrode, colours of materials fade. It is important to use stock within its stock life – also called shelf life. The procedure is called first in, first out – **FIFO**. If you look at the supermarket shelves of perishable items, you will notice that items with the shorter 'use by/sell by' dates are at the front of the shelf to encourage shoppers to take them first.

Product life cycle – this is the concept of the various stages of the product's life until recycling or scrappage. There are regulations referred to as extended producer responsibility – **EPR**. These also cover packaging as well as the actual goods from manufacture to disposal.

Documentation

Spreadsheets – most companies have a variety of spreadsheets for different purposes.

Orders – these are documents sent to the supplier by the purchaser to confirm the intended purchase. The order is a legal document confirming that the buyer will complete the purchase, that is, pay for the goods.

Invoices – these are sent by the seller to the purchaser. They include the VAT. Payment of invoices must be done in an agreed time span. Time spans can vary between immediate, no time allowance and 30 days.

Skills and Questions

1. Investigate the stores or warehouse of your work placement company, ask about the stock management system. If possible get a few days' work experience in this area.
2. Investigate the life cycle of any engineering items, record this graphically with neat sketches.

Chapter 17

Quality Assurance, Control and Improvement

Definitions of the term **quality** are many and varied. In engineering, the term **fit for purpose** perhaps explains it better, especially alongside the term **well made**.

In many engineering fields, the term quality is used to trace the sources of all the materials and components, this is **auditable quality**.

Quality products may be defined by being:

- Aesthetically pleasing
- Reliable
- Having the required functionality
- The appropriate weight
- The ideal size
- Backed up with aftersales service and warranty
- Economic in long-term use
- Resilient to long-term use.

Quality is also about customer expectations; these may be many and varied.

Reliability – this is associated with quality; a quality product is expected to be reliable. Reliability is defined as the mean (average) time between failures. Of course, this is also related to service, maintenance and acceptable use.

Quality control (QC) and quality assurance (QA) – two similar terms often used mistakenly for each other. QC is about checking the item, the product, for compliance with its design and proposed manufacture. QA is about the process by which the component is made – the process.

Quality control checks should be carried out at each of the following stages:

- Design – does it meet the required criteria?
- Purchasing – are the materials of the correct specification?
- Production planning – is this the correct way to make it?
- Manufacture – is the process correct, are the correct tools used?

DOI: 10.1201/9781003284833-17

- Final inspection – are the surfaces within tolerance?
- Despatch – is it packed and moved appropriately?
- Aftercare – is there a correct warranty package?

These are just some of the items that may be checked, it will depend on the product and its intended usage.

Inspection and testing – may take many forms. The usual method is to remove samples from the production line at various points and check them for tolerance and functionality. This relates to Six Sigma, see Chapter 19.

The audit trail – all quality work should be auditable, that is, it should contain information as to where and when, on what component of which batch it was carried out.

Tech note

Audit, a term borrowed from accountancy, meaning an official inspection. In this case documentary evidence from which other documents and actions can be traced, or put into a context, or scenario.

Total Quality Management (TQM) – this is a concept of managing the process of manufacturing – or any other process – completely from customer needs to customer satisfaction. It is a circular activity, one area leading to the next, then repeating the circle. The stages are:

- **Management** – strategic planning, QA/QC policy methods, review stage.
- **Resources** – production planning, staff development needs, design process.
- **Realisation** – manufacturing, quality checks, manufacturing processes.
- **Analysis** – quality procedures, customer satisfaction surveys, usage reports.

ISO 9001 – Quality Management System

See also Chapter 13 on engineering standards.

This standard gives guidance to companies on the setting up of quality assurance systems to an internationally recognised system. It helps to prove and advertise that the company is minded to produce high-quality products. To be accredited to ISO 9001 the company will need to indicate the quality objectives of the business, prepare and produce appropriate documentation, communicate the concepts throughout the business and to customers, carry out audit checks and review meetings with possible action plans.

Legislation – some items must be made to comply with various laws, rules and regulations. An example of this is vehicle emission regulations. Cars, vans and trucks need to meet certain emission standards, that is, their exhaust gases must not contain more than a certain amount of carbon compounds.

Approved testers – there are companies set up to carry out independent testing and inspection to ensure that controlled items meet the requirements of the given standards. These companies are approved by the various government and international bodies to certify compliance. An example is the **Vehicle Certification Agency – VCA.** This is also known as the **UK Vehicle Type Approval Authority.**

Approved suppliers – the materials and components that engineering companies use need to be to a quality standard to meet their needs. They also need delivery and price terms and conditions to ensure their financial viability.

Skills and Questions

1. Write down what the term quality means to you as an engineer.
2. State three reasons why ISO 9001 may be important to an engineering company.
3. What is the difference between quality assurance and quality control?
4. Why would an engineering company only use an approved supplier?
5. Use the Internet to look up the activities of an approved testing organisation such as the VCA and make a list of the services which they offer.

Chapter 18

Principles and Practices of Continuous Improvement

Nothing is permanent except change. These wise words were expressed by the Greek philosopher Heraclitus around 500 BC.

Car manufacturers typically bring out new models every three years, they will update them every year in some way. This is to both improve the vehicle, and make it more attractive to encourage customers to buy new ones. Mobile phone manufacturers may do this on a much more frequent basis – sometimes every few months. The same principles of improvement apply to company management, some companies and larger organisations restructure, or make significant management changes, every three years.

Working in engineering you will soon get used to regular changes, new products, new machinery and new ways of working. Changes are made to:

- Make the product faster
- Make the product better – lighter, stronger, more functionality
- Work in a safer way
- Meet outside influences, including the weather.

Why do we want to continuously look for improvements, let's list some of the reasons:

- Improve the quality of the product
- Reduce manufacturing or service costs
- Make products, or services, more quickly
- Sell more of the product, or services
- Improve safety in the workshop
- Reduce the amount of space needed to complete the task
- Eliminate scrappage
- Reduce the amount of waste.

DOI: 10.1201/9781003284833-18

All these points will generally reduce costs, either directly or indirectly. This can be used to both reduce the price of the products, or services, and increase profitability.

Kaizen is a management tool, or process, which is used to help develop a way of thinking to promote continuous improvement. It is of Japanese origin, 'zen' is the Japanese word for good. There are several variations of Kaizen which are used for different situations, all of them are aimed at positive thinking. Basically, their aim is to get all the staff, yes everyone, to know the facts, particularly the numbers related to the tasks – How many? What size? When? Where? How? Then to ask how can we do it better? If a machine breaks down, ask is there a better machine?

The key aspects of Kaizen are:

> Visualise – tools that will help you visualise what your change process will look like.
>
> Measure – Being consistent is key to making improvements.
>
> Improve – The principles of Kaizen are split into four categories: process, product, people and environment.
>
> Repeat – Repeat the process.

This is a continuous loop, similar to other quality systems.

Kaizen encourages every member of the workforce, from leadership to frontline workers, to propose changes that can improve workflow. The idea is that those workers who are in the 'gemba', or real place, are those more likely to identify real opportunities for improving the flow of their processes. To be successful in Kaizen you must always be looking to eliminate the **eight forms of waste**:

1. Defects: Scrap or products that require rework.
2. Excess processing: Products that must be repaired to satisfy customers' needs.
3. Overproduction: When there are more parts in production than customers are purchasing. This type of waste spells big trouble for an organisation.
4. Waiting: A person or process inaction on the manufacturing line.
5. Inventory: A valuable product or material that is waiting for processing or to be sold.
6. Transportation: Moving a product or material and the costs generated by this process.
7. Moving: Excessive movement of people or machines. It is more common to talk about people movement, as this leads to wasted effort and time.

8. Non-utilised talent: When the management team fails to ensure that all the potential and experience of its people are being used. This is the worst of the eight wastes.

Five main point of Kaizen are:

1. Think of how the new method will work; not how it won't work.
2. Don't accept excuses.
3. Totally deny the status quo, be ready to start new.
4. Don't seek perfection. A 50% implementation rate is fine as long as it is done on the spot.
5. Correct mistakes the moment they are found.

Kanban is a management tool, or process. It translates from Japanese as 'signboard'; it is about visualisation. Engineering is a very visual profession, well-made engineered items are usually aesthetically pleasing, so visualising the work is attractive to engineers. Kanban can be used in most businesses, or areas of work.

There are many ways of using Kanban, the simplest is to use a magnetic board, writing out the steps that the product needs to pass through, then using a magnet and paper to move each product through each stage, timing them as it goes. In some cases this will be like playing the snakes and ladders board game. The snakes are, of course, the problem areas which will be highlighted by products being rejected, sent back for more machining perhaps.

Another illustration of this can be found at airports tracking arrivals, you can find this information on airport websites too. If you choose a major airport and go to arrivals, you will see information actually tracking the passengers of the flights as they pass through the airport. You will see the heading change: Expected – Landed – Customs – Awaiting baggage – Baggage arrived. If you are at the airport your friend will shortly appear after the last message. The airport authorities actually track these times on spreadsheets and use them to put extra staff in these areas, or change the operating methods as needed.

The four main principles of Kanban are:

1. Visualising the work progress – using visual tracking.
2. Limiting the amount of work in progress, that is, starting a job and finishing it, not having the workshop cluttered by work which is not being worked on.
3. Focusing on workflow, ensuring that work in progress flows smoothly, allowing concentration on quality issues.
4. Continuous improvement, by making all the processes clearly visible so that all team members can contribute.

SWOT analysis – this is an abbreviation for strengths, weaknesses, opportunities and threats. Dividing a sheet of paper into four squares, one for each of the headings, a group of people fill in the squares with their views under each of the headings. It is best to do this with a group of people from different departments to get a wider range of views.

Tech note

With any quality activity it is better to have a group of people from a range of jobs or departments to get a wide view of the situation. A group made up of an odd number of people ensures that discussions do not end with a draw – equal each side of an argument.

The results of the SWOT analysis will indicate any actions that are needed, like in Kaizen, you probably don't have the resources to do everything. Pick those that you can do immediately and do them, then make **action plans** for other points which need addressing.

Focus groups – when developing a new product, or a change in service procedures, a focus group made up of customers, and other interested parties, is advised. Car companies will have a few focus groups to review a new model before full-scale manufacturing is started. Restaurants do this by what they call 'soft-opening'; they invite a group of potential customers for a free dinner and ask for feedback to check that both staffing levels and the menu are correct.

Skills and Questions

1. There is a saying that before you can manage other people, you have to be able to manage yourself. There is another saying that if you want to start a new business, the best time to do it is now. You might not be in a position right now to start an engineering company; but you can apply some continuous improvement to yourself. Start with a personal SWOT analysis. See if there are ways in which you can apply the information in this book to improve yourself.
2. Draw up a table to compare Kaizen and Kanban – showing which points are similar and which are different.
3. State three reasons for reducing waste in engineering.
4. State why it is good practice not to have any non-active work in progress in the workshop.
5. Look at any engineered product and suggest three things that could be done to make it better in some way.

Chapter 19

Project Management Principles, Techniques and Practices

Project management is a skill; it requires the efficient and effective use of a number of specialist management tools. These are tools for recording, understanding, analysing and using data.

The skill in project management is getting the right people with the appropriate tools, and the correct materials to the chosen location at the chosen time.

Projects take all sorts of different forms and sizes, for instance, putting the wiring in to a house, making an aircraft, building a skyscraper, or maybe building a garden shed.

Tech note

Gig Economy – Engineering work, in some areas, such as construction engineering, is developing into what is known as a gig economy. That is, more and more engineers make a living by carrying out project work – like musicians performing gigs. This type of work requires project management skills as well as engineering skills. Early completion of projects within budget often leads to large financial bonuses for the project engineers; but the next project will require bidding for, it is not regular weekly work.

PRINCE2 – stands for **PR**ojects **IN** Controlled Environments. It has activities for **starting up, planning, running, controlling and closing a project.** It groups these activities into processes. It is a process-based approach for project management. There are seven processes that are compatible with the Core Principles and guide you through the project.

The seven principles are:

1. Projects must have business justification.
2. Teams should learn from every stage.

DOI: 10.1201/9781003284833-19

3. Roles and responsibilities are clearly defined.
4. Work is planned in stages.
5. Project boards manage by exception.
6. Teams keep a constant focus on quality.
7. The approach is tailored for each project.

There are training courses available to learn the method fully. There are also apprenticeships available in project management.

Six Sigma – this tool gets its name from the six standard deviations which are known by the name of the Greek letter sigma – uppercase Σ or lowercase σ – which make up a normal distribution bell curve. If the bell curve is made up of one million units, or events, there should be no more than 3.4 occurrences of defects.

This is a much-favoured tool and used extensively in engineering manufacturing. Its usual application is when large volumes of the same object are being manufactured, for example toothpaste tubes or food cans, where both accuracy and cost are critical.

There are five steps in six sigma referred to by the acronym **DMAIC**, these are:

1. Define the problem.
2. Measure the performance.
3. Analyse each item.
4. Improve performance.
5. Control the process.

There are specialist training courses for Six Sigma.

Budget planner – whatever you are doing, you need to know what it is going to cost. You need to know what the cost is so that you can know what price you can sell it for, hopefully making a profit.

Tech note

In engineering manufacturing there are two main methods of costing, these are:

Cost plus a percentage and know the typical selling price and make it for less. Sale price less cost of manufacturing equals profit.

The contents of the budget planner may include:

• Materials costs
• Labour costs

- Special machinery use – hire maybe
- Design costs
- Taxes and levies
- Travel and transport costs.

These titles are usually referred to as budget headings – usually just said as *budget heads*.

Budget spreadsheet – this document is used to record the amount of money spent under each budget head. It will show dates and amounts, it is used to give a running account of the money spent against the money available.

Gantt chart – this is the chart by which project activity is both planned and progress reported. On any project it is normal that some jobs cannot be started until others are finished. For example, you cannot put the roof on the garden shed until the sides are assembled. So, this gives you an order of assembly, and the timing of each part of the task. The bigger the job, the more important this becomes. To record progress the actual times are recorded. An important concept with this is **lead time**, the amount of time needed between ordering an item and when it arrives.

The people commissioning the task will normally agree the Gantt chart contents and request updates on progress against the plan.

Milestones – these are the important, and usually clearly visible signs of progress made with a project. Examples of common milestone terms used in engineering are:

- **Manufacturing drawings** – the final approved design, ready to be made.
- **Production prototype** – first actual one made.
- **First fix** – building engineering – all services in place before plastering.
- **Second fix** – building engineering – all services completely installed.

It is normal that money is paid from the budget at the achievement of each milestone.

Tech note

Milestones used to be found on main roads, some are still in existence, to show travellers the distance to the next town or village, this was essential to travellers in the days of horse travel to indicate the distance to the next hostelry (pub) where their horses could be fed. Some pubs are called a halfway house, because they are between two major towns.

Decision tree – sometimes in making decisions a decision tree is useful. This is where you answer yes or no to a question and it leads to another question.

Weighting matrix – when working on a project there are choices to be made, often called options. These options fall into a number of categories. The categories may affect the functionality of a project, and may also have an effect on the costs. So, the various features are given weightings related to how important they are seen to be. Mobile 'phones and their associated packages are a good example. You have choices of:

1. Operating systems.
2. Access to network.
3. Choices in packages of calls, text and data.
4. Size and other physical options.

Which is most important, what is the effect on price, are often questions which are asked.

Risk matrix – project risks are rather like Health and Safety risks, as they can be classified under the two headings of likelihood and severity, or consequences. Both scales are numbered one to five. The value for likelihood is multiplied by the value for consequences. Therefore the lowest value is 1, there is always a risk with any project; and the highest value is 25.

Critical path analysis – this looks at the critical path, or way through the project, which cannot be avoided. It is very much related to the Gantt chart and the milestones. For example when fitting out a building, the building services engineers cannot proceed with a second fix until the plastering is completed and the plasters cannot work until the building services first fix is completed. So there is a critical path of: first fix – plastering – second fix. Sufficient time will be needed for drying after plastering before the building service engineers second fix.

Ishikawa or fishbone diagram – this is a diagram which shows the causes of an event. Sometime expressed as cause and effect. The causes are often referred to as root causes.

PERT – this is an abbreviation for Program Evaluation and Review Techniques. It has five essential steps, these are:

1. Identify the specific activities and milestones.
2. Determine the proper sequence for these activities.
3. Construct a network diagram.
4. Estimate the time required for each activity.
5. Determine the critical path.

Skills and Questions

The best way to learn about project management is to carry out your own project and write a report on it using as many tools and methods as possible. See Chapter 20, *My Project* for further guidance.

Chapter 20

My Project

Most engineering courses entail some form of project work. Some bachelor degrees are 100% about learning by project, and most masters' degrees are mainly focused on projects and the analysis of projects.

Completing a project is about taking a disciplined approach to engineering. In fact, engineering is itself a discipline. That is to say, you must do certain tasks in a particular order, one after the other. Sometimes different tasks can be carried out concurrently, in the design office this is called concurrent engineering.

When you are carrying out a project it really must be something which interests you, projects can become boring if you are not fully committed and interested in the outcome. It also should be something that is new to you, and it may have potential for further learning at the next stage of your education.

Case Studies

A number of students have turned their projects into actual distinguished engineering careers. If you want a career in engineering you need three specific positive things to offer to an employer, these are:

1. Your diploma awarded at the highest level that you can achieve. You diploma is your key to gaining an interview.
2. Work experience. You need this to show that you can start on time and you can work with other people. Social interaction is very important.
3. A skill; you are able to do something that is very specific. This does not need to be at a high level; but something natural that sets you out as an employable individual.

Your project can help you show these points to a future employer. Three examples of projects leading to careers are:

DOI: 10.1201/9781003284833-20

Student 1 – prior to college he had read a lot about engines, he was particularly interested in the Nikola Tesla engine – nothing to do with Tesla cars. The Tesla engine works by using a pressurised fluid to spin disks. For his project he made a Tesla engine out of carbon fibre. In the audience for the project presentations was a senior engineer from a major engine manufacturer. During the lunch break the student was signed up by this company and given a training package. Some of his work is now patented.

Student 2 – he had an artistic aptitude towards engineering and a natural talent for drawing. He went on work experience in the design office of a company that produced gears. His next work experience was at a company which sub-contracted specialist machining work. They were surprised to see his name on the drawings. He could both design and machine gears – both firms offered him a job and training – he now works between the two as a development engineer.

Student 3 – was interested in electronics and computer-based games. He was also very quiet; but diligent. For his project he developed a car dashboard with a range of visual images based on his iPad. He was offered a design post with a major car manufacturer who saw his project – this type of dashboard instrumentation is now used on many cars and motorcycles.

Let's Get Started

1. You need to decided where and how you are going to record it, examples are:
   - Computer – separate file.
   - iPad/tablet – separate file and photos.
   - Phone – photos and notes.
   - Notebook – for jottings and sketches.
   - File folder/wallet – for loose papers.

For projects, like writing this book, I use a notebook, which enables me to make written jottings and drawings in pencil or pen: a file on my computer for digital material and my finished written material – I also copy it to a separate hard drive; a computer file for photographs from my camera and my phone for snaps of items, or information, this I download as needed, also the notes app on my phone and sometimes a short video. Loose papers go into a paper wallet.

When you have made a choice stick to it, it's faster if you have a chosen method, so that you can produce all your evidence at the completion event.

2. Mind map three possible choices – keep the mind map.
3. Using a decision-making weighting table, or grid, and a decision tree make your choice – write a little about this choice. Justify the choice firmly, this will help keep you on track.

4. Read and make notes on ethics and environmental issues.
5. Draw up standard operating procedures (SOPs), where possible.
6. Carry out risk assessments.
7. Build a budget spreadsheet.
8. Draw up a Gantt chart, set of milestones, or time line to estimate the timings to completion.
9. Start to carry out the project.
10. About half way through the project, compare project performance to the planned situation. Asking questions such as: is it safe?, is it on time? and is it on budget? Make suggestions about what you can change, or not change, give reasons.
11. Complete the project.
12. Analyse the result, what went well, what could have been better, how would you do it next time, what would you advise another person to do, how could you develop this further?

Glossary

This section defines a number of the words and phrases used in engineering.

'O' rings	rubber sealing rings
Acceleration	rate of increase of velocity
Adhesion	stickiness
Alignment	position of one item against another
Alloy	mixture of two or more materials; may refer to aluminium alloy, or an alloy of steel and another metal such chromium
Ally	slang for aluminium alloy
Atom	single particle of an element
Bench	working surface; also flow bench and test bench
Beta version	test version of software, or product
Bore	internal diameter of cylinder barrel
Carbon fibre	like glass fibre but used as a very strong carbon-based material
Code reader	reads fault codes in the ECU of the particular system
Composite	material made in two or more layers – usually refers to carbon fibre, may include a honeycomb layer
Condensation	changes from gas to liquid
Contraction	decreases in size
Corrosion	there are many different types of corrosion, oxidation or rusting being the most obvious
Density	relative density also called specific gravity
Diagnostic	equipment connected to the system to find faults
Machine	something that does useful work converting energy into motion
Epoxy	resin material use with glass fibre materials
Evaporation	changes from liquid to gas
Expansion	increases in size
Foam	material used for make seats and other items
Force	mass multiplied by acceleration

Friction	resistance of one material to slide over another
Gelcoat	a resin applied when glass fibre parts are being made – it gives the smooth shiny finish
Glass fibre	light-weight mixture of glass material and resin to make vehicle body
Heat	a form of energy – hotness
Inertia	resistance to change of state of motion – see Newton's laws, inertia of motion and inertia of rest
Kevlar fibre	super-strong material, often used as a composite with carbon
Mass	molecular size, for most purposes the same as weight
Metal fatigue	metal is worn out
Molecule	smallest particle of a material Newton's laws
First law	–A body continues to maintain its state of rest or of uniform motion unless acted upon by an external unbalanced force
Second law	–the force on an object is equal to the mass of the object multiplied by its acceleration (F = Ma)
Third law	–to every action there is an equal and opposite reaction
Oxidation	material attacked by oxygen from the atmosphere; aluminium turns into a white powdery finish, see rust
Parent metal	main metal in an item
Power	work done per unit time, HP, BHP, CV, PS, kW
Prototype	first one made before full production
Rust	oxidation of iron or steel – goes to reddish colour
Spine	backbone like structure
Steward	a senior officer in the organisation of a motor vehicle event
Stress	force divided by cross-sectional area
Stripping	pulling apart
Stroke	distance piston moved between TDC and BDC
Swage	raised section of metal panel
Swage line	raised design line on metal panel
TDC	top dead centre
Temperature	degree of hotness or coldness of a body
Test bench	test equipment mounted on a base unit
Torque	turning moment about a point (Torque = Force × Radius)
Velocity	vector quality of change of position, for most purposes the same as speed

Index